Work Life 2000 Yearbook 1

Springer-Verlag London Ltd.

Work Life 2000
Yearbook 1
1999

The first of a series of Yearbooks in the Work Life 2000 programme,
preparing for the
Work Life 2000 Conference in Malmö 22–25 January 2001,
as part of the Swedish Presidency of the European Union

Organised by:
National Institute for Working Life, Sweden
National Board of Occupational Safety and Health, Sweden
Joint Industrial Safety Council, Sweden

Edited by:
Richard Ennals
Kingston Business School, Kingston University, UK

Springer

Richard Ennals, MA, PGCE
Kingston Business School, Kingston University, Kingston Hill,
Kingston Upon Thames, Surrey. KT2 7LB

Additional material to this book can be downloaded from http://extras.springer.com.

British Library Cataloguing in Publication Data
Worklife 2000 yearbook
  I: 1999. - (AI in society)
  1.Labor 2.Labor - social aspects 3.Employees - Effect of
  technological innovations on 4.Labor laws and legislation
  5.Labor policy
  I.Ennals, Richard
  331
  ISBN 978-1-4471-1227-3   ISBN 978-1-4471-0879-5 (eBook)
  DOI 10.1007/978-1-4471-0879-5
Library of Congress Cataloging-in-Publication Data
A catalog record for this book is available from the Library of Congress

Typesetting: Ian Kingston Editorial Services, Nottingham, England
Printed and bound by Athenæum Press Ltd, Gateshead, Tyne and Wear, England
34/3830-543210 Printed on acid-free paper SPIN 10724703

Work Life 2000
Organising Committee

from the Swedish National Institute for Working Life:
Anders L Johansson, Chair
Bengt Knave, Vice Chair

from the Swedish National Board of Occupational Safety and Health:
Helena Nilsson
Bertil Remaeus

from the Swedish Joint Industrial Safety Council:
Ingvar Söderström
Henrik Lindahl

Secretariat:
Arne Wennberg, Secretary General
Maud Werner

Information Committee:
Lena Skiöld
Ingemar Holmström
Henrik Lindahl

Preface

We live in a fast-changing era. New technologies and the growing flow of information create new conditions for societies and nations as well as for individuals. Politics, economics and business all face dramatically different circumstances. The rate of production is increasing and new technologies and applications are introduced at ever shorter intervals.

Most industrialised countries face the same kind of problems and challenges. New forms of work and enterprise are emerging while many people are totally excluded from the labour market. A rising number of people are working on short- and fixed-term contracts or part-time. Stress and insecurity have increased for many individuals.

At the same time many traditional work environment problems remain. Bad working conditions still affect people's health and women are still often at a clear disadvantage.

In this world of international interdependence, no single national solution is possible. We need a wider perspective and a joint effort. This is the reason for Work Life 2000 that brings to the fore the problems of modern working life.

A Swedish initiative, Work Life 2000 is organised by the National Institute for Working Life, the National Board of Occupational Safety and Health, and the Joint Industrial Safety Council. Through an extensive international process, the most recent research findings concerning working life issues will be compiled prior to the Work Life conference in Malmö, Sweden, which is taking place when Sweden takes over the presidency of the European Union.

The conference, intended for representatives of European governments, government authorities, labour market parties, business organisations and other interested parties, is prepared through more than 60 international workshops led by prominent researchers and experts. The workshops illuminate virtually every aspect of modern working life: the labour market and labour law, the work environment, the information society, work organisation and small and medium-sized enterprises.

After each workshop a scientific summary is produced, as well as a popular/journalistic summary available by subscription. In your hand now you have another product: the first Work Life 2000 yearbook, a detailed account of what have been discussed during the different workshops. It gives you the possibility to enter deeply in the material and, by help of the index and the comments, see the connections between the different themes.

However, Work Life 2000 is not just a gathering of knowledge and information. Equally important are the processes that start at each workshop. Dialogues across the nations get on their way, new networks are being built – and this is something that will go on long after the conference in January 2001 is closed.

Anders L. Johansson
Director-General
Swedish National Institute for Working Life

Helena Nilsson
Director-General
National Board of Occupational Safety and Health

Ingvar Söderström
Managing Director
Joint Industrial Safety Council

December 1998

Contents

Indexes 209

1

Introduction to Work Life 2000
Workshops

The first Work Life Yearbook, Work Life 1999, arises from the preparatory phase of the Work Life 2000 Programme, and covers the period from December 1997 to December 1998. As the first of the series of three Work Life Yearbooks, it lays the foundations on which the following two years of work are to be built. Participants at the Work Life 2000 Conference in Malmö in January 2001 will receive the set of three Yearbooks.

The Context for Work Life 2000

Changes in science, technology and the global economy are transforming the nature of working life. As we approach the Millennium, the Swedish National Institute for Working Life has embarked on a major programme of research and workshops, in association with the National Board of Occupational Safety and Health and the Joint Industrial Safety Council.

The Work Life 2000 Conference in Malmö, from 22–25 January 2001, will be a major European event at the start of the Swedish Presidency of the European Union. The Conference is part of an ongoing process of European social dialogue, supported by the Swedish Government and Social Partners. This includes a series of over 60 international preparatory Workshops, which started with the first Work Life 2000 Workshop in London in December 1997. Issues involving Working Life and Work Environment have been identified as a major theme for 2001, and as an ongoing agenda focus for the newly expanded European Union.

Two overall themes will characterise the contents of the Conference and Workshops:

- Job Development and Creation: Labour Market Strategies

- The Good Working Life: Work Environment and Organisation

The Conference is thus not an isolated event, an objective in itself, but a public reflection of the Work Life 2000 process, and of developing European policy.

The design of the Conference will be consistent with that of the three Yearbooks, to be published by Springer-Verlag, which arise from the Work Life 2000 workshops. The set of Yearbooks prepares participants for the Conference, and offers a dissemination route for the Workshops, which have resulted in agendas for future work.

The Workshops are organised in five themes, which run through the three Yearbooks, and which will be presented at the Conference.

1. Labour Market

2. Work Organisation

3. Work Environment

4. Small and Medium Sized Enterprises

5. Information Society

The Workshops have typically been held at the office of the Swedish Trade Unions in Brussels, involving 15–24 invited participants over two or three days. Others have been held at venues around Europe, in London, Copenhagen, Amsterdam and Dublin. The approach has been based on dialogue, rather than on formal presentations. Participants have come from all European Member States, from researchers, governments and the Social Partners, as well as from European Union institutions, with some invited non-European experts. There have been popular journalistic reports of each Workshop, and scientific reports from the Workshop organisers, published by the Swedish National Institute for Working Life.

The Malmö Conference is for about 650 invited decision makers in the European Member States, drawn from EU Member State governments, relevant public agencies, labour market parties, business organisations and other interested parties. It is intended to reflect the atmosphere of the Workshops, while requiring less specialist technical expertise. The proceedings and conclusions of the Workshops will be made available through the Yearbooks and popular summaries to those who attend the Malmö conference. The conference sessions will be concerned to set agendas for action during the period of the Swedish presidency and beyond.

Thus the Yearbooks are seen as a means of communication between the expert groups who have attended the Workshops, and with the decision-makers attending the Conference. It is intended that the set of Yearbooks should also be published on CD-ROM in time for the conference, with hypertext links between workshop themes. In each volume the indexes will facilitate cross-referencing.

Work Life 2000 Workshops

The workshop series commenced in 1997, laying the foundations for the Malmö Conference in 2001, and developed its own distinctive patterns. The first Yearbook draws on the following:

Labour Market

1.   Transnational Trade Union Rights in the European Union

Led by Niklas Bruun, and held in London, 15–16 December 1997.

2.   Rights at Work in a Global Economy

Led by Niklas Bruun and Brian Bercusson, and held in Brussels 8–10 June 1998.

Work Organisation

1.   Ageing of the Workforce

Led by Åsa Kilböm, and held in Brussels on 23–24 March 1998.

2.   Psychosocial Factors at Work

Led by Eva Vingård and Töres Theorell, and held in Copenhagen on 24–26 August 1998, in association with the International Conference on Occupational Health, with the same title.

3.   Managing and Accounting for Human Capital

Led by Jan-Erik Gröjer and Ulf Johanson, and held in Brussels 14–16 September 1998.

4.   Occupational Health and Safety Management Systems and Workplace Change

Led by Kaj Frick, and held in Amsterdam 21–24 September 1998.

5.   Developing Work and Quality Improvement Strategies

Led by Jörgen Eklund and Bo Bergman, and held in Brussels 19–21 November 1998.

Work Environment

1.   Space Design for Production and Work

Led by Jesper Steen, and held in Brussels 20–22 April 1998.

2.   How does Medical Surveillance contribute to the Objectives of Framework Directive 391/89?

Led by Anders Englund, and held in Brussels 7–9 September 1998.

3.   Environmental Management and Health and Safety

Led by Henrik Litske, and held in Dublin 3–4 December 1998.

4.   Occupational Trauma – Measurement, Intervention and Control

Led by Tore J. Larsson, and held in Brussels 14–16 December 1998.

Small and Medium Sized Enterprises

The series of Workshops in this theme begins in 1999. Innovative Networking of Small and Medium Sized Enterprises (SMEs) is addressed as a theme in research in Work Organisation. The new universities in Sweden have taken on the Third Task of working on local and regional economic development, which includes working with SMEs.

Information Society

1.   Research Dissemination

Led by Elisabeth Lagerlöf and Markku Aaltonen, and held in Brussels 23–25 November 1998.

The effective use of Information and Communication Technologies underpins the Work Life 2000 Programme.

General Themes and Conclusions from the First Year of Work Life 2000

1.   Cross-References and Inter-Relations of Themes

The workshop programme is organised in themes, but there have been numerous connections and cross-references between themes. The journalist and Yearbook editor have highlighted points of contact, which may lead to future projects and collaborations.

2. The Importance of the Process of Dialogue

Work Life 2000 has established a distinctive dialogue process, which will be continued through to the Conference. Invited participants contribute to discussions without lengthy formal presentations. A common language is established, and practical measures are identified by which action plans can be implemented. The dialogue is inclusive and expanding, and the process is often as important as the product.

3. Workshops as a Means of Strengthening International Networks

Invited participants come from across the European Union and beyond, to attend meetings which both reinforce existing collaborations and help build new networks. Because participants are not strictly representative of their nations or organisations, the discussion is able to move beyond official positions, finding new ways of characterising situations in a way that may facilitate policy development. This enables an authoritative group to find ways around what have previously been obstacles.

4. The Emergence of a European Approach to Productivity and Innovation

There will be no one simple recipe for European economic success, but the dialogue process is highlighting the nature of differences within Europe, and the ways in which diversity can be managed for collaborative competitive advantage. The Workshops provide an opportunity to listen to the experience of others. Sweden has very distinctive experience to contribute, but it is clear that Southern European Member States, in particular, have had different traditions. As a first step, we are finding a common language to describe these differences, and a partnership framework in which it can be taken forward.

5. Action Research as a Critical Resource

The pace of change is such that conventional research methods cannot keep up. One valuable contribution from Sweden has been the robust tradition of action research, including large scale national programmes and evaluation studies. There have been similar traditions in other Northern European countries. As the European Union seeks ways of formulating and testing new policies, the availability of action research case studies is extremely important, together with the expertise of action researchers.

Dialogues on Working Life

In order to try to capture the atmosphere of the Work Life 2000 workshops in this Yearbook, they are reported as a series of dialogues, introduced by the organisers. These were not policy-making meetings, or occasions for voting. Although there

were participants from across the European Union, they were not formally representative of their nations or organisations. Within the workshops there were few
extended presentations. In some cases papers had been prepared and circulated in
advance, sometimes making use of Web sites. The workshop time was devoted to
discussion, in an environment that encouraged exploration of new ideas.

Each workshop was planned and prepared separately, but as part of the overall Work
Life 2000 programme. Workshop leaders are preparing summary reports of the
deliberations, and scientific reports incorporating the full text of selected papers,
with the goal of supporting the Malmö conference. There is also a series of popular
reports of each workshop, prepared by the journalist Lena Skiöld. These incorporate
additional material from interviews with leading participants.

Each workshop brought together expert practitioners and researchers in specialist
areas, who could be seen as sharing a vocabulary and a set of concepts. Each brought
a different perspective, meaning that precise definitions were rare. The objective was
not to secure agreement on particular policies or action plans, but to establish a
process of European discourse leading to the conference.

What is the Work Life 2000 Dialogue about?

The series of workshops address different aspects of Working Life. The areas under
discussion are not separate, and indeed there are numerous overlaps. The same
subjects are considered by different professional groups, thus producing related
discourses. The reader may wish to use the index for cross references, and the facilities of the CD-ROM which will accompany the completed set of three volumes.

The set of dialogues is each about working life, based on experience of practice. With
few exceptions, each participant has attended only one workshop. The main exceptions are journalist Lena Skiöld and rapporteur Richard Ennals. Whereas Lena
Skiöld has offered popular external accounts of the separate workshops, Richard
Ennals has tried to identify the vocabulary and concepts underlying each workshop.
His objective has been to participate in the dialogue, and to present a written
account that can be accepted by the participants of each particular workshop, and
then, after further editing, be made accessible to the Work Life 2000 community as a
whole, as well as to a wider audience.

There have been some common patterns in the conduct of the workshops, especially
those held at the offices of the Swedish Trades Unions in Brussels. Invited participants arrive from all over Europe, take a seat around the table, and introduce themselves in turn. The workshop organisers have been responsible for setting the
agenda, issuing the invitations, and coordinating the preparation and circulation of
papers in advance. They have determined the sequence of speakers and interventions in the discussion. The Work Life 2000 secretariat have provided unobtrusive
administrative support, and coordinated catering arrangements.

Each workshop dialogue has its own distinct characteristics, but some general
trends can be identified:

- The chairman is not also the main speaker. In some cases there have been two
  workshop organisers, taking the chair alternately to enable the other to speak.

- Speakers have been discouraged from "grandstanding", giving pre-prepared formal presentations. The workshop environment has favoured discussion rather than lectures.

- Equal respect has been shown to all participants and contributions, with a view to learning from different experience, and from systems in different European member states.

- There has been interest in case studies, which cast light on differences.

- The desired outcome of each workshop has been the identification of a framework for ongoing dialogue in the field.

Approaches to Dialogue: the Role of Theatre

During two conferences in Stockholm in October 1998, "Dialogues on the Sciences and Humanities", which formed part of the celebrations of Stockholm European City of Culture 1998, a number of different approaches to dialogue were presented.

In the first, "Philosophical Dialogues", the orientation was primarily philosophical, working from original texts or translating earlier dialogues. The interesting exception was the work of Martha Nussbaum, a leading feminist philosopher, whose dialogue was based on a book chapter she had written about her emotions on the death of her parents. Playing the leading role on stage, she was delivering her own words based on her own experience.

In the second, "Dialogues on Performing Knowledge", the focus was on dialogue between science and humanities. The opening example was the work of John Monk, a Professor of Technology, who has ventured into writing dialogue within the genre of Zamyatin, Orwell and Huxley's novels on the future.

At each of the conferences, those who wanted a rigorous dialogical form were often disappointed, and saw the work under consideration as more conversational than dialogical, more literary than philosophical, and at time simply personal. Those who had agreed to comment after presentations found that their prepared texts took them in diverse directions, and it was difficult or impossible to sustain dialogue in the subsequent discussions with the audience.

By contrast, discussion in the Work Life 2000 workshops has involved contributors from within the same set of professional cultures, using purposive means of communication to which they are accustomed. The dialogue has a past and a future as well as a present: it offers us a snapshot of the form of life. It is not simply cold and rational, but is an arena in which emotion can be expressed and action engendered.

One conclusion is that professionals with shared experience of working life often have more in common than academics from different disciplines who have been brought together to consider the topic of dialogue, which is not fully defined. This has practical implications for the future of Europe, as it suggests that common practical experience can overcome diverse backgrounds, as long as we continue to value our differences as a source of strength and competitive advantage.

We do not need to arrive at a single rigorous dialogue that combines all the many layers of sub-dialogue represented in the workshops. We may observe that words are used differently in the different professional groups, which in some cases have enjoyed traditional rivalry. It is important that we establish that it is possible for an outsider to learn to understand the specialist dialogues to the extent of being able to contribute and "know how to go on". In the Work Life 2000 workshops, the rapporteur starts as an outsider.

This lesson from working life can be carried back to the intellectual world of "Philosophical Dialogues", where different modes of discourse can inhibit cross-disciplinary communication. It may remind us why philosophers such as Wittgenstein appealed to examples from ordinary language and working life, and why the Swedish approach to Practical Philosophy has found such international appeal.

The Work Life 2000 Conference will benefit from the integral involvement of theatre and performing arts in the furtherance of the European Dialogue. In 1998 Stockholm has been European City of Culture. When in 2001 Sweden holds the European Presidency, that holistic approach should be preserved. At the 1998 British Presidency Conference on Work Organisation, held in Glasgow, Forum Theatre was introduced into otherwise conventional conference proceedings, with a presentation by British and Danish actors with experience of using theatre as a tool for organisational change. Ongoing work with regional development coalitions in Sweden, Italy and Germany, described in the supplementary materials, is using Forum Theatre as one means of addressing the language barrier that otherwise impedes wider participation in European Social Dialogue.

The central actors in Malmö will be those involved in working life, and with responsibility for making decisions about the future of working life. Malmö will be the stage, and all the men and women in it merely players. Close by, across the new bridge, will be Kronborg Castle, legendary home of Shakespeare's Hamlet, Prince of Denmark. At the "Psychosocial Factors at Work" dinner at the castle, part of the Work Life 2000 programme, participants were left in no doubt as to the relevance of theatre for the workplace. Hamlet faced the challenges and stresses of an unsustainable set of relationships, and used theatre to "catch the conscience of the King".

One view of the Work Life 2000 workshop process is that it has been a reversed form of Shakespearean playwriting. The invited actors come together, and participate in active dialogue, organised by the leaders in a sequence of sessions. The rapporteur seeks to derive the script from which they may have been speaking. Before they depart, the actors are able to amend the draft script, so that it reflects what they intended to say. The dialogue continues by electronic and other means, and the script of the workshop report is a by-product, which has collective ownership. It is then taken as raw material from which the edited Yearbook is produced. Readers may choose to identify the characters, and form a view as to whether there is a consistent plot.

Conclusions

Dialogue offers a way forward, taking account of the exchange of ideas, new information, external developments and the passage of time. Around the table in a single room, great progress can be made, on a pre-announced topic, by those who come prepared, and with prior acquaintance with some of the others present. Dialogue is a process, offering benefits from participation whatever the outcomes or conclusions.

In practical terms, it can be hard to provide direct involvement for large numbers of participants. Personal dialogue underpins subsequent electronic communication.

What happens outside the workshop room? How do trade unionists talk to employers? Do they go through national or industry sector routes? How do they involve others who have not been present at the workshop dialogue? How do we overcome the various language barriers?

How do the different professionals come to terms with the complexities of Europe? For example:

- In the field of occupational health, how do we deal with the different relationships between doctors, nurses and social work professionals in each European country?

- In the field of human resource accounting, how do we make sense of diverse accounting and human resource management processes?

- A European Directive may consist of few words, but involve considerable complexity and ingenuity if the requirements are to be met across the European Union.

The process of dialogue is fundamental to European democracy. We may be redis-covering some of the core issues of practical democracy, this time through the consideration of working life. The Work Life 2000 workshops should be seen in this context. They represent an ongoing European dialogue process, initiated from Sweden, conducted in English, and involving participants from all Member States of the European Union.

Dialogue provides a point of contact between cultures. More information on each workshop, including full texts of many of the papers presented, can be obtained from the workshop leaders. The ongoing dialogue offers Europe a development organisation in the field of working life, a virtual international research institute. Participants come from different structures, sectors and nations, and are able to join a developmental dialogue process that explores different interpretations of the past and present, and opens new opportunities for the future. The Office of the Swedish Trade Unions in Brussels is an ideal location for such work.

2

Workshop Proceedings

Labour Market

1. Transnational Trade Union Rights in the European Union

The workshop leader was Niklas Bruun, of the Swedish National Institute for Working Life and Hanken School of Economics, Helsinki, Finland. The workshop was held at the Offices of the Trades Union Congress, London, 15–16 December 1997.

Abstract

Transnational trade union rights have moved to the top of the agenda in the EU, with the imminent adhesion of the UK to the Social Chapter of the Maastricht Treaty and its incorporation into the EC Treaty at the Amsterdam Summit in June 1997. The issues concerned are multiple and complex.

1. The legal framework for the EU social dialogue, which has produced the Parental Leave Directive and an Agreement on Part-Time Work.

2. The legal implications of the growth of European Works Councils, with an estimated 400 agreements in multinational enterprises and another 800 to be negotiated over the next three years.

3. The law governing transnational industrial action, following disputes in the French and Spanish transport industries and the closure of the Renault plant at Vilvorde.

The development of an EU strategy for transnational trade union rights at EU level is a central priority. Speakers had circulated papers in advance.

Welcome

Anders L. Johansson, Director-General, Swedish National Institute for Working Life, introduced the first workshop. Labour Law was a key issue with which to start, with major cases being decided at the time of the workshop.

John Monks, General Secretary, Trades Union Congress, set the scene in terms of European Union Presidencies from the UK in 1998 to Sweden in 2001. The workshop is in preparation for the Swedish presidency in 2001: this is advanced planning; the UK presidency started two weeks later in January 1998, with preparations still to be made!

The context of the single market and single currency will mean deeper integration. Enlargement to the East means radical change. Cross-border capital transfers and mergers will increase. Two recent disputes – the Renault plant closure in Belgium, and the French lorry drivers dispute – have had a major impact. The European Court has considered such issues, with implications for the right to strike. The new Social Chapter will provide a new single basis for social policy: possibly a revolutionary change. The UK still has domestic problems over regulation and workers' rights, with arguments about job creation. There is a need for an active social dimension in Europe. Regression in social legislation, levelling down, must be opposed. All social provisions must go to the social partners under the social protocol. Increased negotiation and the increased role of the European Parliament offer prospects of improvement.

FUNDAMENTAL SOCIAL RIGHTS AND EQUAL OPPORTUNITIES

Fundamental Rights and the Outcome of the Inter-Governmental Conference

Professor Lammy Betten, University of Exeter, argued that we need a legal basis for freedom of association legislation, yet rights to strike etc. are excluded from the text of the Amsterdam Treaty. As yet social partners and member governments have agreed no legislation on right to strike, but stopgap provisions are not a sufficient basis for legislation. At present freedoms are included in community law, but as a basis for litigation, not legislation. There are human rights provisions to protect trade union rights, but a community bill of rights would not necessarily underpin transnational trade union rights. On the key conflict of principles between freedom of movement and the right to strike: it was not clear whether the regulation would be annulled.

Klaus Lörcher, Legal Adviser, Deutsche Post Gewirtschaft, shared the view that positive legislation is unrealistic at present, and that the views expressed by European Commissioner Monti, setting competition against the right to strike, pose new challenges. Litigation is the way forward. The European Social Charter will need to have more impact on legislation and litigation on social rights in the community. **Lammy Betten** called for a short-term strategy by the European Trades Union Congress on litigation, concerning the European Court of Justice and European Court of Human Rights. We should not forget the medium and long-term strategy regarding fundamental social rights: Trades Unions must address this politically.

Regan Scott, Transport and General Workers Union, London, asked about case law on trades unions as legal persons, which is a relevant issue for European litigation. **Brian Bercusson** indicated that such issues would arise first at national level.

Antoine Jacobs raised issues from **Lammy Betten's** paper, where she highlighted odd exclusions from the Amsterdam Treaty, probably because employers and unions dislike intervention by legislators. Can there be intervention by other means, using other articles? He thought the European Court would find ways. Surely the Council and Commission are busy legislating about social partners, positive legislation which would be open to attack. **Erland Olauson,** of the Swedish trade union confederation LO, explained the omission of fundamental trade union rights from the treaty. Legal industrial action would have to be defined, which has not been done in a number of countries, where the principle has been that anything is allowed unless it is prohibited. The imposition of European positive law in the Swedish system would break it down, so Sweden, Denmark and Finland were not enthusiastic.

Erik Carlslund gave a further account of the negotiations on behalf of the European Trades Union Congress. Positive rights were kept minimal, but case law is being developed. This involves a clash between freedom of movement and the freedom to strike. **Lammy Betten** added the clash with the free movement of workers (a human right), which should be seen as a different right from the free movement of goods (an economic principle). There are two concepts of fundamentality. Would a corridor approach violate the right to strike or limit its application?

Dave Feickert of the TUC raised different concepts of obstacles; this is a practical problem for truck drivers, which will intensify with deregulation in 1998. These issues are becoming complex and real. **Brian Bercusson** considered case law on free movement, noting efforts by the European Court to remove obstacles to free movement. **Lammy Betten** had argued that Amsterdam excluded legislation, but there is scope for social dialogue agreements. Community law will continue to develop: what is the role for agreements?

Niklas Bruun commented on restrictions on the right to strike, and the Monti proposal on free movement. This is to protect property rights, not human rights. **Erland Olauson** argued against legislation on fundamental rights, and asked about the implications for the European Court of Human Rights. **Lammy Betten** responded. The exclusion of legislation, because it creates rules, is unimpressive. Legislation will come through case law. At a certain point, you must have rules. If something is excluded from the treaty, it will be harder to ground in law. She notes that the International Labour Organisation accept arguments in terms of economic well-being.

Klaus Lörcher cited the Gustafsson case and the French case, noting the use of litigation by employers, with the objective of reducing trade union rights. **Erik Carlslund** cited the Convention on Human Rights and the role of social partners in modern democracy. Where does this leave trade union rights? When considering enlargement, what account would be taken of ILO definitions and transnational issues? **Lammy Betten** noted the problem of representativeness. A democracy includes trade union rights, but institutional details vary. Who do the unions represent? People, workers, the unemployed? Transnational trade union rights will be a matter of legislation, not merely litigation. It is a question of prohibiting via legislation, or fighting via litigation.

Brian Bercusson considered the coming five years, planning the next set of objectives for the Inter-Governmental Conference. What do we do in the interim? **Erland Olauson** favoured blocking new legislation. Perhaps we need a more proactive strategy of social dialogue agreements. The issue becomes one of strength of agreements, some of which are not legally binding. This can be done on a sectoral basis. Norway, Denmark and Sweden have experience of Basic Agreements.

Equal Opportunities and Transnational Trade Union Rights

Professor Linda Dickens, University of Warwick, asked whether there has been enough on equal opportunities already. Some past limitations have been overcome. The debate has broadened to consider families and work. However, women continue to be disadvantaged, in the workplace, in the EU, and among social partners. Renewed emphasis on social dialogue means rights of unions and obligations of social partners come under scrutiny, in terms of internal equality. Could collective bargaining at a transnational level be a preferable means of enhancing equal opportunities? Top-down measures tend to be more male-oriented. Realities are changing, with the feminisation of the European labour market. Collective bargaining could be enhanced by taking on equality issues. Trades union legitimacy is enhanced if unions are seen as representing all workers. On the other hand, partial collective bargaining can diminish equality.

Transnational collective bargaining could be used to promote equal opportunities, but this requires change. Women's proportional presence concerns internal and external equality. Women tend to be under-represented on European Works Councils. Experts have not included those concerned with equal opportunities, and a gender perspective tends to be absent. As for the European sectoral dialogue, she has been analysing collective agreements across the European Union, with **Brian Bercusson**, and noting openings for sectoral social dialogue in different member states. There is uneasy cohabitation between trades union rights and equal opportunities.

Jo Morris, Equal Opportunities Officer, TUC, argued that the key issue is internal union democracy. There is an opportunity for transformation of the roles of trades unions: equal rights in the workplace imply questions about internal democracy. Unions must be seen to be representative if they are to flourish, and gender representation is a prerequisite.

Collective bargaining could prompt a change, from sectional interests of equal opportunities to mainstream, from moral imperative to integrated action. This requires structures to empower women at each stage: the medium is the message. She cited the new European directive on part-time working, and noted a gender divide. She welcomed the idea of a mandatory requirement for proportional representation of women, but noted that unions could make such changes themselves.

Hans Flüger responded regarding European Works Councils, which are widespread in the metalworking industries. There have been many missed opportunities, but proportional representation of women could have meant reductions in their numbers. He noted the omission of provision for Central and Eastern Europe, for tactical reasons. These were not issues to die for. **Jan Cremers** took a similar line, on behalf of building and wood workers. They have secured a smaller proportion of agreements, and there was pressure of time, and other priorities dominated. Equal opportunities could feature in future revisions. The issue needs to be addressed first at a national level, and in the workplace.

Erik Carlslund raised practical questions, suggesting female quotas in delegations. However, if it is a matter of choosing a single representative, that is difficult. It may be excessive to determine both what is to be agreed and by whom. He cited the directive on part-time working. **Brian Bercusson** was more provocative. Neither politicians nor trade unions had given priority to equal opportunities. Is there a case for further litigation? Should agreements be challenged if not gender equal? Could this threat enhance the chance of change?

Dave Feickert spoke of experience of recent agreements, deriving from work in equal opportunities. The Commission has had difficulty making equal opportunities a priority. Progress has been made in the European Parliament. National laws vary. **Erik Carlslund** quoted the prohibitions on discrimination, from the text of the directive on part-time working. **Linda Dickens** talked of mainstreaming and gender-proofing, and saw the potential for legal challenges. She presented equal opportunities as an opportunity to be grasped, and not just one of a list of problems to be dealt with. **Jo Morris** argued against just leaving equal opportunities to evolve: progress needs demand and political will. Equality is a process by which we can embrace the world in a new way.

Antoine Jacobs argued that social partners act as social legislators: it could be argued that delegations should be balanced. Do we support a prescription to social partners? Is this a legal possibility? **Erik Carlslund** noted that this is an issue for employers as well as trades unions. **Margit Wallsten** noted that employers do not claim to be democratic in basis: a start has to be made at the bottom, at enterprise level. Change can be made from within. **Jan Cremers** noted that negotiations are undertaken by the organisations and members concerned. There are problems of balance between workplaces and nations which were seen as more important than gender.

Lammy Betten noted that the employers' representation at the conference was wholly female, but that females were in a minority among trades union representatives. She noted that Parliament, Council of Ministers and Court of Justice have gender imbalances. Active recruitment of women to trades unions would help with problems of representativeness. We need to establish that we have understood what women want.

EUROPEAN UNION SOCIAL DIALOGUE

Trade Union Rights

Professor Brian Bercusson, University of Manchester, highlighted key issues. Actors participate, with rules based on processes, and outcomes emerge. Transnational rights have an impact at national level. Rights can be seen at three levels: association, autonomy, and action.

Association

There is no one European legal position for trades unions at present. The rights of association at European level are held by organisations, not by workers. Thus legitimacy at European level depends on national level organisation.

Autonomy

The European Commission has to identify social partners for social dialogue: thus far they have chosen ETUC. There are issues of autonomy. ETUC votes on key issues, and national delegations may be outvoted.

Action

Problems arise through consultation with the Commission. Is the right to be consulted sufficient? Should the social partners be obliged to enter dialogue? How strong should transnational trade union rights be? One problem can be the absence of employer organisations at sectoral level with whom to negotiate. There is a lack of resources. How can the ETUC and others cope? The dialogue on part-time work was hard to handle.

What are the legal effects of the agreements, on those who negotiated, on national organisations, and on individual employers? It is easy when the agreement becomes a directive, but cases will come when implementation is a matter for member states. Are member states to be required to implement?

Erland Olauson, LandsOrganisationen, Stockholm, set out the reasons for trades unions to participate in the social dialogue: to improve the conditions of members and ensure that decisions made by the EU in the area of working life fit with their views. This gives the EU a social dimension and a broader legitimacy. The Commission will listen if it is clear that ETUC have support, but most ETUC members do not know the dialogue exists. We need to take the local perspective. They need to know of possibilities of participation, cooperation across borders, and who will pay. The European Works Council directive concerns only private sector, and affects relatively few women. Members must see concrete results emerging. Agreements at a European level are not legally binding through the ETUC, but only if endorsed by the Council. The objective is to improve working conditions, giving legitimacy to the process. Perhaps we have to start with legally binding agreements at national level. This could mean rights to participate at national level, rights to contact similar organisations across Europe, assurances that costs will be borne by governments and employers, assurances that sympathy action can be taken.

Staffan Olsson from DG-V considered the court case, due in spring 1998, regarding an employer who declined to participate in social dialogue. **Hans Flüger** asked why the discussion is necessary. It could only be because the national level game is ending, and there are fears about potential anarchy in arrangements in other countries spreading to the formerly smoothly run countries. There are concerns arising from globalisation of private industry. National regulatory regulations are not up to the task. Managers have gained power at the expense of workers. Even if we have to discuss the need for basic rights, why not coordinate bargaining voluntarily in a globalised economy? We could agree, for example, 1752 hours per year. Some would say this was not legal, others would say they were not strong enough, and support could be mobilised. There needs to be regulation of teleworking. Regulation is seen as politically dangerous.

Dave Feickert talked of south, north and deregulated west. The balance of experience and tradition flows into ETUC. The mixture of traditions is helpful. Theoretical legal issues need to be answered in practical negotiations. Amsterdam has brought into fruition the political objectives of trades union leaders. We now need good negotiators. **Bruno Veneziani** of Bari University argued that the key issue concerns collective bargaining. He saw social dialogue as fluid, best handled without the interference of law. Implementation of European agreements may restrict autonomy. Practical details would vary nation by nation.

Lammy Betten dealt with the legal definition of ETUC. Members of national trades unions would need to be consulted regarding their representation by ETUC, leading to binding agreements. This has implications for national law. Without resolving such matters European collective agreements will be hard to implement. **Jan Cremers** saw the reality of ETUC's legal current status. European negotiators have been mandated to negotiate on behalf of workers, and have signed agreements. The only useful aspect of the cocktail of traditions is the resulting instrumental approach. Legislation can achieve more than social dialogue.

Brian Bercusson felt that there are limits to what can be achieved at national level, and thus scope for transnational legal rights as a means of achieving progress at national level. Benefits will vary between nations. **Erland Olauson** is convinced that European cooperation is necessary. The question is how we should do it. In the Nordic nations there has been a century of experience of buoyant trades union membership and success. What can the Nordic model contribute to the stabilisation of a new European approach?

John Foster of the British National Union of Journalists indicated that his union would benefit from progress in transnational trade union rights. Five groups control the press in the UK, in an attack on unions and Europe. He described a current case of de-recognition at Associated Newspapers. Mrs Thatcher had supported free trades unions in Poland, but not in the UK. He described the sequence of legal stages in the UK, leading to a case at the European Court of Human Rights. There is a right to be a member of a union, but no right to be represented. The new Labour Government is committed to reintroducing rights, but may be more likely to follow the recommendations of the CBI than the TUC. A European approach to transnational trade union rights is vital. **Erik Carlslund** noted that the European Parliament is discussing such matters at present. Some progress can be made by the trades unions concerned. He reflected on the implications of recent summit statements on vocational training and work organisation, and the involvement of the social partners.

Regan Scott discussed derogations from the working time directive, and rights of individuals. He then considered the implications of the political block which had been applied to the apparently available legal basis for trades union rights.

Jan Cremers saw Europe as a means of reassessing the value of national bargaining systems. The cocktail keeps changing. The Nordic model was successful for a century, but the question is what will be appropriate today. We all tend to stick to our national solutions, making it difficult to cooperate.

The Role of European Union Institutions

Professor Antoine Jacobs, University of Tijlburg, is sceptical about social dialogue. It fits in with the state as spectacle, something to show on television. The EC can help stimulate social dialogue by sending more questions, but this raises questions of practicality and resources: secretaries, interpreters, information, training of negotiators.... Major investment would be needed. If not successful at one level, then dialogue must be conducted at another level, such as via sectors.

The European Parliament does not want to promote the social dialogue, as the subject matter is beyond their influence. Perhaps the social partners should pass draft agreements to the European Parliament for amendment. The European Court will not feel obliged to support or stimulate social dialogue, but will be involved in the process, interpreting the contents of agreements. We might prefer a specialist body. The Economic and Social Committee has a limited role. Social dialogue, both formal and informal, will be involved.

Peter Morris, UNISON, argued that the system is in flux, and we can see increased integration of European economies, requiring responses from trades unions. The ETUC took its present form in the 1980s, and linked with the Catholic and social networks under the Delors Presidency of the European Union. The ETUC were critical insiders, leading to the 1991 Social Agreement, a pragmatic agreement to get around the roadblock of social legislation. We have moved from networking to institutions, and unions need to be able to deal with employers and governments. The ETUC has a good research secretariat, but remains small in face of large challenges. The consideration of part-time working posed difficulties, and resources are stretched. The resources are insufficiently integrated. There are questions of derogation to specialist union groups. There will also be sectoral negotiations with cross-sectoral implications (e.g. the deregulation of energy). The key issue is national and European, facing UNISON and others today.

Kirsten Precht of HK/Industri in Denmark described the approach taken in Denmark, seeking to add provisions concerning union rights to the treaty, giving clarity to the legal status of collective agreements. The social partners are given the tools to implement and enforce agreements. The role of ETUC should become one of monitoring, adding to the efficiency and effectiveness of trades union responses. This is not simply an export of the Danish model, but builds on prior European experience.

Hans Flüger noted the absence of interest by employers in joining dialogue for general benefit, while trades unions and wage earners learn who is in charge. He argued the case for regulation and political pressure. He depicted UNICE as likely to explode if obliged to organise around a positive campaign. Are results from sectoral dialogue creeping back into national bargaining? He cited the recent agreement in the food sector, with an agreement including remarkable legal features. If there are no benefits at the national level, why take it seriously? Negotiation is about involving partners. What matters is the legitimacy and representativity of the partners. He preferred the European to the Anglo-American model. Defensive renationalisation of negotiation is not enough: we need a new European system.

Jan Cremers could envisage a role for the European Parliament. **Brian Bercusson** saw the role of the European Parliament in applying pressure on employers. A new constitutional settlement is required. The Parliament has resources. If UNICE is not capable of delivering results, then sectoral partners are needed. **Margit Wallsten**, from UNICE, noted that there will be a heavy burden of negotiation, with both sides seeking to benefit those they represent. She sees little role for the European Parliament. She noted that UNICE has limited resources. Possibly in future negotiation will be at a European, rather than national, level. At present, UNICE can only really deal with one issue at a time, and do not seek resources from the Commission. Negotiation experience takes time to accumulate.

Bo Rönngren argued that negotiation skills will develop over the coming months, with a series of tasks to be undertaken. The Social Summit might not have been the best way of handling issues concerned with employment and the labour market. The answer is not legislation and institutions, but developing voluntary social dialogue. **Antoine Jacobs** noted that politicians like to refer problems to the social dialogue. There can be more than new rules and new agreements: in addition the dialogue can monitor existing rules. What is happening on unemployment, gender inequality, etc. The workload could become very heavy, but resources are limited.

EUROPEAN WORKS COUNCILS/TRANSNATIONAL INDUSTRIAL ACTION

EWCs: The Lessons of Renault

Professor Marie-Ange Moreau, University of Aix en Provence, highlighted lessons, noting important legal ambiguities and complexities. The case demonstrates the value of litigation by trades unions. European directives to date have made insufficient provision for transnational issues, and we need to consider national and European legal interpretations. Problems remain when seeking to sue transnational employers. Information rights raise new issues, concerning time and decision making. European Works Councils and trades unions have the right to context management. Mergers and re-structuring problems transcend borders, and raise major problems.

Hans Flüger, General Secretary, European Metalworkers Federation, argued that legal approaches are no substitute for political change. There is insecurity regarding the working of European Works Councils, which can only be done at a European level. He was glad that his union had won the cases in both Nanterre and Versailles, but wondered how a German court would have decided. The case showed the limits of information and consultation. The Renault case has influenced company practices: Electrolux engaged in consultation, but probably without changing the eventual outcome. The same is true of ABB. The financial press are being used as means of influencing the stock markets, with reports of proposed major redundancies. The law on consultation remains unclear, and there has been no challenge to the business decisions behind the Renault decision to close the plant. The directive on EWCs is inadequate, as has been argued by EMF and ETUC, but UNICE were unable to respond. We must expect further delays and legal mutations. Multinational companies have responded to an organised trade union position. EWCs are a parallel, not an alternative, trade union body. He expressed some doubt about the real calibre of modern management, despite the rhetoric of globalisation.

Patrick Itschert spoke of the Levi's agreement, in the context of likely company restructuring and relocation. Trade unionists were melted down with representatives of other workers, placed in a minority, and obliged to develop a rival structure. Levi's announced major closures and redundancies in the USA, and major closures are expected in Europe, yet capacity is increasing in Turkey, Pakistan, Hungary and elsewhere. Trade unions refused to agree the Works Council agreement, and have set up a meeting with management. The legal aspects need clarification, including the status of the trade union groups. Pressure led to trade union recognition in Hungary, and growing support.

Jan Cremers spoke about the effect of the Renault case. There is now more respect for formalities in France, but this is largely window-dressing. What if the case had arisen in another country, based on different law? He has seen no real effect on British or German management. He seeks more consistency. Trade union ideas have been streamlined. As a result of restructuring in the building products industry, there is a case for using Works Councils as platforms, as instruments for social dialogue. **Staffan Olsson** reported that the European Works Council directive had been extended to the UK the previous day.

Hans Flüger responded that the employment conclusions of the Social Summit commit the Commission to establishing structures to deal with restructuring in industries such as metalworking. This was a result of pressure by ETUC. Who will sit on the new panel? Commissioner Bangemann is no friend of trades unions. In a recent meeting of the car industry, themes of globalisation and benchmarking were preferred to employment. Employers should not be given the veto.

Brian Bercusson considered legislation, dialogue and litigation. Little can be done on legislation, though there are proposals at national level. Many EWC agreements would now be phrased differently, but revision is not due in the near future. Dialogue could accelerate the process. On litigation, the question is where to take proceedings. It is a question of choosing the right jurisdiction. In some countries, the EWC directive can be used as a tool to improve the situation.

Hans Flüger reflected on experience with employers such as Honda, with useful results from EWC based work. In metalworking there are 600 EWC in action, and coordination is difficult. Litigation adds to the pressure. There is some ambiguity about the role of enterprise-based trade union activity. Change through constitutional means is fine when you win cases, but less enjoyable when you lose.

Jan Cremers identified three stages in the quality of agreements related to new directives:

1.   early pioneers of a new communication channel could gain useful results in a context of trust, with much not written down;

2.   then there is a transition phase, given the presence of the directive, with the motivation of voluntary agreements, based on what is written but without mutual positive approaches, and with a minority trying to act separately;

3.   finally negotiation is tough, and it is hard to go beyond the minimum.

Marie-Ange Moreau noted that management are transnational, and the workers must respond. She cited the Hoover case in 1993, when a transnational trade union position was not developed. The Renault case changes the range of tactics available.

Legal Problems of Transnational Industrial Action

Niklas Bruun concentrated on free movement of goods and industrial action, the subject of recent documents and cases. He took the case of French lorry drivers, and referred to documents that were circulated, including the 9 December 1997 judgement of the European Court. The EC Treaty requires free movement of goods, and national governments have responsibility: thus private individuals can claim damages. He then referred to the proposal for a Commission intervention mechanism in order to eliminate certain obstacles to trade, but without affecting the exercise of fundamental rights recognised under national law. Under the proposal, the Commission can define an obstacle, and set a short time-scale within which nation states are obliged to respond. The confusion of industrial action with kinds of illegal action is unhelpful.

He referred to the Nordic tradition of strike neutrality. It is not clear that the European Commission have kept to such traditions. The risk has been passed to the member states. He did not expect the regulation to pass in the European Council, where unanimity would be necessary. The problem remains, with decisions in the hands of the Court. He noted a subsidiarity paradox.

Erik Carlslund, ETUC, argued for the primacy of the right to strike. There are relevant cases at national level, but not yet at EU level. The Danish Prime Minister has defended the right to strike against EU restrictions. Do we want legislation or case law? We are in a single market, but fifteen separate labour markets. It is left to national courts to determine whether cases fall under EU rules. Trades unions need to decide how they wish to proceed. Collective bargaining will be affected. He cited debate in the German parliament concerning liberalisation of postal services, and the UPS case. In a single market, with a single currency, separate labour markets are less viable. This is a joint project with employers.

Uno Westerlund noted that the ETUC in 1996 discussed similar issues, concerning the kind of input to be made to the IGC. The outcome was relatively disappointing. The Luxembourg Jobs Summit considered the Social Charter and the Social Chapter. What input will ETUC have in future? We need a European labour relations system that deals with the internal market, and possibilities of transnational action.

Lammy Betten noted that the linkage of the court case and the regulation suggests a dangerous form of interference by the Commission, giving precedence to free movement of goods over the right to strike. The issue was criminal activity, not the right to strike. Legality is determined by national law.

General Discussion

Susanne Christensen has been working on EC directives. She was responding to the draft directive tabled that morning, and recommended that we take account of ILO views. The Danish government intervenes if there are social consequences of strikes.

Bo Rönngren reflected that there are EU rules, and ETUC rules, concerning what should be negotiated at the European level, and what at national level. The issue of being overruled by a majority in the ETUC is relevant: if ETUC does not deal with negotiated legislation, someone else will do so. However, if this occurs in the case of wages, it would be more serious. Equalisation of working conditions, for example, raises questions. The ETUC constitution was amended in 1995, allowing for negotiations and time-scales. There are now new challenges in employment policy, and the social partners need to consider a new set of rules of the game. He was surprised that the EU proposed directive has gone so far without discussion with the social partners, and asked about the role of DG-V. He noted a lack of confidence in the EU from ordinary people. **Brian Bercusson** agreed that the coordinating role of DG-V appeared to have failed.

Erland Olauson spoke about the right to strike, and the right of association. There is no problem at national level with the right to strike, but the European level is hard, due to the range of national systems. How, then, do we respond to suggestions of restrictions? Perhaps the treaty should include principles to be respected by the European Union when exercising European law. This may be better than over-reacting to court cases. **Brian Bercusson** argued that such a treaty change could enable us to take the Commission to the Court, but the Court cannot be referred to itself.

Pekka Aro, of the ILO in Brussels, cited the body of law on freedom of association and the right to strike. ILO conventions set out principles of tripartite consultation. There are definitions of what constitute workers' representatives, and their rights. It is unfortunate that European discussions omit the ILO context. ILO is happy to cooperate. There is no need to reinvent the concepts at a European level.

Klaus Lörcher discussed the case law of the European Court of Justice: it was not enough to disregard cases and move to the political agenda. Lawyers for employers will study these cases, so we must consider alternative strategies in terms of case law. He referred to the European Social Charter and the European Single Act, making reference to the European Convention on Human Rights. However, the question of transnational trade union strikes has not been dealt with thoroughly, and work will need to be done in advising governments and securing appropriate outcomes.

Jan Cremers was not surprised at the lack of consultation on the draft proposal. He wondered whether anything had really changed in DG-V and beyond after Amsterdam: only DG-V cares about trades unions, while the other Directorates have little interest, arguing that these are technical matters. We need fine tuning of proce-dures within ETUC, concerning social policy and collective bargaining. How do we get mandates from unions, in the context of European negotiations with tight time-scales and limited communication within Brussels Directorates-General?

Kirsten Precht was concerned with transnational agreements, with transnational parties at a European level. Court systems will need to be uniform. In the short term the parties could decide how to resolve disputes.

Kirsti Palanko-Laaka talked from Finnish experience. There are worries about the proposal from Commissioner Monti, apparently prepared in secret without consul-tation. The Finnish government respects the right to strike and tripartite negotia-tions, and objects to violations of the right to strike. The struggle will continue. European trades unions should be active. The British Presidency will include the discussion of the new proposal.

Brian Bercusson noted that unanimity is required for the proposal. Commissioners, including Padraig Flynn, appear to have let the matter through.

Kieran Connolly of the NUJ discussed General Strikes, which are not protected under British law. An infringement of trade union rights in Europe could justify a European General Strike. How would the legal authorities respond? **Brian Bercusson** noted that the Court ruling concerns illegal acts: in the UK, with restrictive legisla-tion, this could include many strikes.

Hans Flüger noted the search for pragmatic responses, and the yearning for some unanimous principles. This involves writing constitutions at a European level, and would require a consensus. The Nordic unions oppose federal approaches. The Germans are not present, but assume German standards must prevail. Defensive rejection of Monti's proposal is not enough. The current treaties were phrased to avoid clarity, but hard issues need to be addressed, core to our democracy. The right to strike is being treated as a technical issue in the context of mobility of goods. Multinational companies have transcended the nation state as a regulatory unit, and this requires a functional answer. There is no viable easy constitutional fix. Denmark learned this the hard way.

Thomas Blanke responded from Germany. It is difficult to teach about European Labour Law and single issue directives. There is no logic or precise structure, but a series of initiatives by Commissioners: this is no way to build a legal system. We have four freedoms, four values, but no fundamental rights. Can we build markets without a welfare system and social rights? We need precise rules, and a European constitution. The absence of such a process, supported by the trades unions, is probably because of the principle of subsidiarity, based on national power, itself undermined by globalisation.

Staffan Olsson advocated litigation from social partners following the Amsterdam Treaty. The Employment Guidelines, agreed at the recent summit, called for social partners to be involved: perhaps they could take the lead, becoming involved in employment policy, where the Council depends on them. There are new possibilities. **Brian Bercusson** noted the open invitation to the social partners to join the table.

Bruno Veneziani noted that the Italian constitution gives an individual the right to strike, which is managed by the union.

Antoine Jacobs considered the idea of a European system of law, but noted that Europe is built from pieces. Litigation following Monti will first be national, with damages from the French government. Labour lawyers need to be prepared to rethink their national labour law. The Dutch would say there is no problem: the lorry drivers were not striking, but blockading, which is illegal.

Carola Fischback-Pyttel, from the European Public Services Union, asked whether the social actors are fully transnational? Following the agreements on part-time work, there are new challenges of coordinated responses. National unions have an easier task than transnational groups. How do you achieve a mandate? An *ad hoc* position may have problems. Time limits present problems of resources: for example, working in five or more languages.

Klaus Lörcher argued that European case law will be important. There is no one German perspective. He noted the principle of most favourable conditions in labour law, allowing for progress.

Bo Rönngren noted ways in which affiliated organisations can take part in ETUC, with anomalous structures of federations. It would be difficult to extend time limits. The problem is following up European agreements, with inadequate translations.

Geoff Thomas argued that British labour law is appalling. He feared that New Labour may well take the CBI line on recognition. In the UK litigation will not achieve much: we need to derive benefit from Europe. The trades unions should demand a change in legislative frameworks. There are more opportunities in Europe at present than in the UK for trades unions. Those present have the opportunity to make change.

Anders L. Johansson addressed the issue of promoting the social dialogue, which will become more complex. The problem is not students, but governments and lawyers: there will be disagreements over rules. We know about legal systems built up by bureaucratic means, which lack legitimacy. The real task is to clean up the complexity. Case law can be used like corks in a leaking boat, but the system needs to be more transparent. Key strategic issues include unemployment: a major threat to relations between employers and trade unions. The Nordic model was not about labour law in a technical sense, but about regulating supply and demand in the workforce. Explanations are not to be found in the labour law system itself, but in the balance of power between the social partners. We need research on successful regulation of demand and output of workers and the labour force, with a more market basis: such solutions are to be found. Complexity is a market opportunity for lawyers, while trades unions have the weakest position, as they have to argue their case to members, for whom the proof of the pudding is in the eating. Members need to see something in it for them. Research is needed on demand and output of labour as such, looking at the market rather than labour law.

PANEL: FOCUS ON SECTORAL LEVEL?

Some Key Questions

Professor Brian Bercusson, University of Manchester, asked what do we do after Amsterdam? Do we wait five more years, or are there things we can do? Are there key transnational rights, or key sectors? Can progress be made through social dialogue? Can we use equal opportunities as the basis for sectoral social dialogue? How do we prioritise issues for the social dialogue? What are the resource issues in sectoral social dialogue? What is the way forward for European Works Councils, singly and collectively? With regard to transnational industrial action: should we be sector specific?

Hans Flüger of EMF argued that the impact of monetary union has not dawned on national unions. European expansion will have major impacts. Globalisation will place pressure on employment, wage-earners and unions. Regulatory power must be focused at international level. It would be good to seek to define rights as if from the start. Practical innovation will take place at the sectoral level. Harmonisation of minimum standards is necessary, to avoid social dumping. Not all of Europe has to participate. The role of the Community needs to be clarified. There is a struggle for the soul of Europe.

Eric Carlslund of ETUC wanted to clarify concepts of social dialogue. There is a phase for action and for reaction. ETUC are calling for the protection of trade union representatives in transnational bodies. There must be mutual recognition of social partners. The key issue is the follow-up to social dialogue and the Social Protocol, including access to vocational training and industrial change. Are the social partners ready? They are also concerned with part-time working. The sectoral level will be important in future. We may expect to be regulated through sectoral agreements. Core rights should have been in the treaty and established at European level.

Margit Wallsten, of UNICE, was the only employer representative present. Employers have yet to address these issues. Employers argue for a more flexible labour market, with the minimum of regulations, but globalisation requires a wider perspective on competitiveness. She was not sure that UNICE could cope with sectoral negotiations. It is a confederation of national groups, with sectoral organisation. She argued for caution in the use of concepts such as social dialogue at the

European level. Social dialogue is a curious amalgam of cooperation and negotiation.

Jan Cremers of EFBWW noted the increased concreteness of the single market. Negotiations are either centralised or localised, and not at sector level, which is often ignored at European Commission level. There have been joint statements with employers on health and safety in construction. There are no real practical problems about being legal social entities. He has been pleading for a better division of labour with sector federations. Works Councils can be a platform for social dialogue.

Patrick Itschert of ETUC, Textiles, Clothing and Leather talked about his members. They see 100,000 redundancies per year, and cannot afford further Renault cases. The issue is jobs. They cannot wait for five years, or for further revisions of ETUC and UNICE agreements. The priority is sectoral social dialogue. Employers must address social issues, such as illegal workers. Why not the same concerns for security in textiles as in agriculture? There is also concern for transnational rights, including in China and Nepal, where there may be forced and child labour, and a code of conduct has been agreed. Works Councils are important, with 26 agreements in 1996, and 2 in 1997.

Carola Fischback-Pyttel of EPSU asked about social dialogue in the public services. It may enable us to reach a European industrial relations system, and standards. Any issue can be sectoral: for example part-time work and working time. The sector is not normally seen as transnational, but is a large employment area, and highly feminised. There are different concepts and traditions: public services and civil service etc. There are no representative employers associations at transnational level. Links are being established with electricity companies. Regional government is another huge employment area, and some joint statements have been reached. New structures are emerging, for example in Denmark, as employers and unions form negotiating cartels for particular purposes, while preserving independence of activity. What will monetary union mean for collective bargaining and wage bargaining?

Pekka O. Aro of ILO argued that discussion of industrial relations at a European level needs a wider frame of reference, as unions need to work with colleagues globally. Globalisation is affecting more areas of work in Europe, and has impact beyond Europe. Regional industrial relations structures must take account of Central and Eastern Europe, and Russia. Hungary, Czech Republic and Poland are becoming integrated, but this needs to become part of current discussion. Governments in these countries have not been friendly to trade unions. Trade union membership is falling. ILO conventions are old, but sound. Times have changed: the context for the conventions was industrial. There will be opportunities to extend the conventions in 1998 when discussing contracting. He supports regional conventions and agreements, and some conventions still need ratification.

Staffan Olsson of DG-V identified a growing pattern of agreements resulting from social dialogue. There is nothing to stop progress on sectoral agreements, which could then be offered for consultation with other sectors, possibly eventually becoming directives of the Commission. **Brian Bercusson** welcomed the idea of extending agreements across a sector. **Antoine Jacobs** noted that employers are often not interested in sectoral dialogue, and that there have been complaints about Directorates-General not being interested in the social dimension and sectoral dialogue. Pressure on Directors-General could help motivate employers.

CONCLUDING SESSION: THE RESEARCH AGENDA

The Research Agenda

Monica Breidensjo, TCO, Stockholm, represents TCO on the Board of NIWL. The workshop had been fruitful and open, providing a good basis for the Work Life 2000 conference planning. Little had been said on equal opportunities after the paper on the first day, but new research would be undertaken. The issues raised demand cooperation between unions and researchers. We need to be more proactive, and less reactive. The dialogue must continue, using the network that now exists.

Niklas Bruun reflected on risks in academic exercises in European labour law, constructing complicated legal frameworks and problems. Dialogue with users of these tools is extremely valuable. There are dangers in running with the latest cases and the latest Commission papers. Academics have a longer-term responsibility, to explore the interference between national and international levels in labour law, and the influences of each on the other. We need to consider relations with the social dimension and economic rights, and to conduct ongoing comparative research in European labour law. Social dialogue, works councils, and transnational action remain key themes. Work needs to be done on the implementation of directives, and labour law aspects of public procurement. The Employment Chapter in the Amsterdam Treaty needs attention.

Workshop Participants

Juri Aaltonen, Union of Finnish Theatre Employees, Finland
Kerstin Ahlberg, NIWL, Stockholm, Sweden
Pekka Aro, ILO, Brussels
Brian Bercusson, Manchester University, UK
Lammy Betten, Exeter University, UK
Thomas Blanke, University of Oldenburg, Germany
Monica Breidensjo, TCO, Stockholm, Sweden
Danny Brennan, DG-V, Brussels
Niklas Bruun, NIWL, Stockholm, Sweden
Eric Carlslund, ETUC, Brussels
Peter Carrick, Amugruppen, Sweden
Stephen Cavaliers, Thompsons Solicitors, UK
Susanne Christensen, Ministry of Labour, Denmark
Nick Clark, TUC, UK
Penny Clark, ETUC, Brussels
Stefan Clauaert, ETUC, Brussels
Kieran Connolly, National Union of Journalists, UK
Jan Cremers, European Federation of Building and Wood Workers, Brussels
Linda Dickens, Warwick University, UK
Jan-Erik Dolvik, Arena, Norway
Richard Ennals, Kingston University, UK
Dave Feickert, TUC, London
Carola Fischback-Pyttel, European Public Services Union, Brussels
Hans Flüger, European Metalworkers Federation, Brussels
John Foster, National Union of Journalists, UK
Anita Halpin, National Union of Journalists, UK
Ingemar Hamskär, TCO, Sweden

Petra Herzfeld-Olsson, NIWL, Sweden
Reiner Hoffman, ETUC, Brussels
Patrick Itschert, ETUC, Textiles, Clothing and Leather, Brussels
Antoine Jacobs, Tijlburg University, Netherlands
Anders L. Johansson, NIWL, Stockholm, Sweden
David Johnsson, NIWL, Sweden
Carolyn Jones, Institute of Employment Rights, UK
Reinhard Kuhlman, European Metalworkers Federation, Brussels
Katri Linna, LO, Sweden
Klaus Lörcher, Deutsche Post Gewirtschaft, Germany
Jonas Malmberg, NIWL, Sweden
Marie-Ange Moreau, University of Aix en Provence, France
Jo Morris, TUC, London, UK
Peter Morris, UNISON, London, UK
Pamela Morton, National Union of Journalists, UK
Stig Norgaard, Ministry of Labour, Denmark
Erland Olauson, LO, Sweden
Staffan Olsson, DG-V, Brussels
Kirsti Palanko-Laaka, SAK, Finland
Kirsten Precht, HK/Industri, Denmark
Kaija Rantala, Finnish Metalworkers Union, Finland
Bo Rönngren, LO, Sweden
Regan Scott, TGWU, London, UK
Bo Sjögren, Swedish EU Office, Sweden
Lena Skiöld, NIWL, Sweden
Sven Svensson, LO, Brussels
Geoff Thomas, TUC, London
Bruno Veneziani, Bari University, Italy
Margit Wallsten, UNICE, Brussels
Uno Westerlund, TCO, Sweden
Steve Whittington, DTI, UK
Jonathan Zeff, DTI, UK

Reflections on the Workshop

This was the first workshop in the series, held in the Council Room of the TUC in London, with a large attendance that inhibited detailed dialogue. The contributors included strong representation from trades unions, employers, from the European Commission and ILO, as well as academic researchers. It was a genuinely educational occasion, in that major cases in European labour law were under consideration by the European Court during the workshop, which brought the debate alive. It was held in the early months of the New Labour government in the United Kingdom, and introduced by the General Secretary of the Trades Union Congress. Participating British trade unionists had more questions than answers. There was no doubt as to the topicality or the practical importance of the discussion, and a growing awareness that the defence of trade union rights requires coordination at a European level.

A fuller account of the workshop is available in "Transnational Trade Union Rights in the European Union", edited by Petra Herzfeld Olsson, Brian Bercusson and Niklas Bruun, published by the Department of Labour Market Research at the Swedish National Institute for Working Life, Report No. 36, 1998.

2.   Rights at Work in a Global Economy

The workshop leaders were Niklas Bruun and Brian Bercusson. The workshop was held at the Office of the Swedish Trade Unions, Brussels, 8–9 June 1998.

Based on a report by Lena Skiöld

Abstract

The EU is the world's largest economy and trading bloc. In many Member States there are longstanding policies imposing basic minimum labour standards on employers through social clauses in contracts for the supply of goods and services or the performance of works. The international dimension of such a policy is reflected in the ILO Convention No. 94 of 1949 on Labour Clauses (Public Contracts) and a number of EU Directives on public procurement.

In the USA social clauses have been used as positive measures in order to promote the position of minority groups. Some multinational companies are also using social clauses in their contracts with suppliers. Within the World Trade Organisation (WTO) there is an intense debate on the incorporation of basic labour standards in social clauses in trade agreements.

The workshop aimed to build on experience at Member State, EU and international level to develop proposals for the EU to adopt a policy of social clauses in trade agreements.

Working conditions and pay rates vary widely throughout the world. As world trade expands and the economy becomes more and more global, the question becomes more urgent: should we secure the same fundamental rights for everyone who works? And, if so, how?

Discussion

Improving working conditions has a value in itself, but there are other aspects to consider. For example, we have long been aware that there is a link between working conditions and competitiveness. The basic assumption is that improved conditions, in the form of minimum wages and reduced working hours, for example, may bring higher costs to businesses. As a result, according to **David Johnsson** and **Jonas Malmberg** of the National Institute for Working Life, it has been traditional for people wanting to improve working conditions in their own country to seek agreement with others on a common standard. This lessens the chances of being out-competed.

Today the debate on rights at work is taking place, above all, between mature industrialised countries and developing economies. In the West, many people are afraid that the globalised economy will lead to a "race to the bottom", meaning that countries will compete for the lowest standard. The developing countries, on the other hand, view demands for uniform, agreed rights as a form of protectionism. As they see it, the agenda of the industrialised countries is, above all, to protect themselves from competition, and social rights linked to work will have to develop as the economic growth allows.

If consumers want to strive only to buy goods produced under acceptable conditions: for example, not by children; several options are open.

International Systems of Regulations

International systems of regulations have been built up by two important world organisations, the UN's International Labour Organisation (ILO) and the World Trade Organisation (WTO).

ILO

ILO, which includes representatives of governments, trade unions and employers, adopts conventions and recommendations on working life and the associated fundamental rights. These conventions then have to be ratified by the member States in order to become binding. However, the scope for ILO to impose sanctions is small, and not all member states ratify all conventions.

WTO

WTO (the World Trade Organisation) lays down legal and institutional frameworks for international trade. Nearly all the countries in the world are members. The main objective of WTO is trade without discrimination. The cornerstone here is the "most-favoured nation" principle, whereby products from all member states must be treated with the same amount of favour as the products from any one of them. As a result, to treat a country differently because of the working conditions prevailing there is difficult. Decisions at WTO are, as a rule, required to be unanimous. Because many of the member states are against clauses on social rights for employees, the prospects of using WTO to link such rights with international trade are not very bright.

Brian Bercusson argued that rights at work are part of fundamental human rights, as defined for example in the UN declaration on human rights. For example, if you use children or slaves as workers, you are violating their human rights, which weigh more heavily than the most-favoured-nation principle. This line of reasoning also brings us away from the view of labour simply as a cost item.

There are also other methods that are used to promote good working conditions. Agreements, for example, may be used to offer extra benefits to countries where fundamental rights at work are respected, and these benefits may also be withdrawn if they are not. Social labelling, an equivalent to "Environment approval marking", is another option that may be used to offer consumers a choice. Voluntary sectoral agreements, or codes of conduct, also exist. The IKEA company has agreed one with the International Federation of Building and Wood Workers.

The European Trade Union Confederation for Textiles, Clothing and Leather and the employers' organisation Eurotex have also reached an accord on a code of conduct. The code covers 60–70 per cent of companies in the industry in Europe, and prohibits forced labour, child labour and discrimination. It also guarantees freedom of association and the right of negotiation.

Patrick Itschert, Chairman of the European Trade Union Confederation for Textiles, Clothing and Leather argued against accepting exemptions for "cultural" reasons. Child labour is child labour. In his industry, between 100,000 and 200,000 jobs have disappeared in Europe since the early 1990s, of which 15–20 per cent was directly attributable to companies' moving production to other locations. He predicted the same trend in other industries as well, pointing out that you just have to go to Eastern Europe to see that companies offer employees considerably poorer conditions than they do in Western Europe. The next step for textile workers, according to **Patrick Itschert**, is to bring the code of conduct into national collective bargaining agreements. In that way, the code of conduct is combined with traditional labour law. Then different codes of conduct could be harmonized and extended to other sectors. **Patrick Itschert** is fairly optimistic. He believes in this model. If we achieve many measures of this kind, it will change negative attitudes. Companies are becoming increasingly interested in joining up. For competitive reasons, they want the same standard to apply to all. However, it is important for the EU to reach an agreement.

The EU has so far been sending out conflicting signals. Free competition is the mainstay of the Union, but on occasion the traditional policy of competition conflicts with social objectives.

The European Union and Links with Other Countries

Allan Rosas, of the European Commission Legal Service, saw scarcely any obstacles that prevent the EU from also considering social rights as fundamental human rights. On the other hand, the EU itself cannot ratify the Council of Europe's Convention on human rights, or any ILO conventions. The European Court in Luxembourg has concluded that such actions lie outside the formal authority of the Union. **Allan Rosas** argued that this makes it difficult to claim that countries that the EU trades with must ratify the conventions, but it does not mean that the Union cannot promote human rights.

A few examples:

The EU May Include Social Clauses in General Trade and Cooperation Agreements

This is the case, for example, in agreements with Mexico and Cambodia. The clauses refer to fundamental human rights, so it is implicit that certain rights at work fall into this category. If the rights are not respected, the agreement may be suspended.

The EU May Also Take Human Rights Into Consideration Through its Own System of Regulations

There, too, reference is made to general principles on human rights, rather than specific conventions. If a country is considered not to fulfil the requirements, beneficial conditions may be withdrawn. This happened in the case of Burma, which used forced labour. At the same time, the EU sued the State of Massachusetts in the USA on the basis of the State's decision to prohibit its agencies from trading with companies active in Burma. **Professor Christopher McCrudden**, of Oxford University, observed that this was a somewhat contradictory course of action.

When the EU is Offering Financial Support to a Country, the Union Has Considerable Scope for Employing Social Clauses

Allan Rosas explored the restrictions imposed by WTO. How can countries be treated differently without coming into conflict with the most-favoured-nation principle? If the countries concerned contravene fundamental international law it is feasible to withdraw favourable conditions, but where should we draw the line? Nuclear tests?

Social Dumping Within the EU

The situation within the EU is interesting, especially as the danger of social dumping, according to certain economists, is greatest in areas where countries are at about the same economic level. There is an EU directive laying down that Member States are required to ensure that foreign workers sent to another country work under the same conditions as those applying to nationals under the law. The directive may take effect at the same time as the ILO Convention 94 and the directive on public procurement in the EU. Each has different possibilities for imposing sanctions. **Niklas Bruun** confirmed that the legal situation is very unclear. Here, we need action at EU level.

Public Procurement

The directive on public procurement is also a subject of discussion. The question is: can procuring government agencies base their actions on anything other than commercial criteria when choosing among tenders? Yes, said **Professor Ruth Nielsen** of the Copenhagen School of Economics. Others took a different view, but she pointed out that, for example, tenderers lacking serious intent may be ruled out and that companies must meet professional and financial conditions, in order to be included in the tendering process. She argued that there may be scope for criteria that are non-commercial. Take, for example, a case where a tenderer has dismissed personnel in an illegal manner. It should be possible to exclude such persons.

Professor Kai Krüger, of Bergen University, Norway, concluded that it is acceptable to strive for social goals, but that public procurement should focus on economic aspects, and that we should not deviate from the principle that the decision-making process should be easy to follow.

In summing up, **Niklas Bruun** confirmed that globalisation had created the need for new instruments to protect people's rights at work. This is true at all levels: national, EU, and global. **William B Gould IV**, of the American National Labour Relations Board, observed that there are a number of possibilities, meaning national laws, EU legislation and the possibility to negotiate collectively via the European works councils in the EU.

Several delegates maintained that ILO ought to have a stronger position. **Auke Haagsma** of DG-XV argued that ILO should pass international laws that are implemented and policed nationally. **Patrick Itschert** noted that everyone accepts that there are police everywhere, except in world trade. ILO needs sharper teeth.

Workshop Participants

Brian Bercusson, Manchester University, UK
Niklas Bruun, NIWL Stockholm, Sweden
Jan Cremers, European Federation of Building and Woodworkers, Brussels
Carola Fischback-Pyttel, European Public Services Union, Brussels
Hans Flüger, European Metalworkers Federation, Brussels
William B Gould IV, American National Labour Relations Board, USA
Auke Haagsma, DG-XV, Brussels
Ingemar Hamskär, TCO, Sweden
Patrick Itschert, ETUC Textiles, Clothing and Leather, Brussels
David Johnsson, NIWL, Stockholm, Sweden
Kai Krüger, Bergen University, Norway
Lena Maier, NIWL, Stockholm, Sweden
Jonas Malmberg, NIWL, Stockholm, Sweden
Christopher McCrudden, Oxford University, UK
Ruth Nielsen, Copenhagen School of Economics, Denmark
Alan Rosas, European Commission, Brussels
Lena Skiöld, NIWL, Sweden
Erika Szyszcack, Nottingham University, UK

Reflections on the Workshop

This workshop took a truly global perspective, seeking to apply the values of trade union rights to the complex area of world trade. It built on the dialogue at the previous seminar in the series, which had some of the same participants. Issues were raised regarding the role and efficacy of international bodies such as the ILO.

Work Organisation

1. Ageing of the Workforce

The workshop was led by Åsa Kilböm, Swedish National Institute for Working Life, and held at the Office of the Swedish Trade Unions, Brussels, 23–24 March 1998.

Abstract

The aim of the workshop was to identify problems encountered in working life by the elderly workforce, to describe the magnitude of the problems, and to suggest suitable actions to be undertaken by governments, at the workplaces, and by the individual. The purpose of such actions is to reduce premature retirements, and improve working conditions and health as well as the work ability of the elderly workforce. These goals may be reached through improved work environments, better training and education programmes, information about the capacities of the elderly, and legislation, particularly concerning early retirement and rehabilitation.

Middle-aged and elderly people constitute a growing proportion of the labour force, in parallel with the demographic changes, i.e. ageing of the entire population, in most developed countries of the world. Employees above age 50 have a very high rate of premature retirement, many of them have health problems especially of the musculo-skeletal and cardio-vascular systems, and they have difficulties finding new jobs when faced with unemployment. In addition, preconceptions about the abilities of elderly people are common. Negative expectations occur regarding their ability to learn and adjust to new situations, and with regard to health and physical capacity. Positive expectations refer to loyalty to the employer, experience and stability. As a consequence of predominantly negative expectations, age discrimination is common, e.g. in the EU administration.

During the economic recession at the beginning of the 1990s, governments handled this situation mainly by restricting retirement and sick-leave benefits. Such actions tend to be blunt, and they do not consider the possibilities of improving the fit between the employee and his/her work. The workshop focused on three issues: the labour market situation for the older workforce, their ability to learn and develop new skills, and health problems encountered. For each issue possible remedies were identified, and activities undertaken in different countries to improve the situation were outlined.

Peter Warr, Institute for Work Psychology, Sheffield University
Age, Competence and Learning at Work

Peter Warr spoke as a research psychologist, but gave particular attention to policy implications. He based the presentation on eight questions:

Are Older Workers Less Good than Younger Workers?

There is no overall difference, but there are variations between jobs. Information is lacking about specific job features. The better people (or the worst) may have left already. We need long-term studies, looking at entry and exit. Older employees are more consistent, with work that is slower but of higher quality; they are less likely to leave, and have fewer accidents and non-sickness absences. Overall, there is no difference in absentee rates. Behaviour and financial indicators showed no overall difference in costs. Training costs are lower at higher ages, when staff turnover is reduced at those ages. Older staff are initially hesitant about new technology, but are soon competent. Older customers like older workers.

Why Might One Expect Older Workers to Be Less Good?

Physical mechanisms wear out, so we expect the same of bodies and minds. We are aware of memory failures at older ages, but there are age stereotypes and prejudice. Laboratory studies of information processing show poorer performance by older people in selective attention, and memory of previously presented material. Of central importance are declines in fluid intelligence, working memory and information processing speeds. These decrements are present within the working population, from the age of 20. Longitudinal studies are needed, but there is a problem of differential exit from work. Laboratory research tests the limits of workers' abilities, while much of everyday life is more straightforward.

Does Not Experience Help Older Workers?

People can learn, unlike physical machines. Experience (activities over time) is to be distinguished from expertise (knowledge, wisdom and skill). Years of job experience can correlate with job effectiveness (up to 5 or 10 years). Where older people are seen as better at jobs, experience has been shown to be important. Expertise is typically greater at older ages: for example job knowledge. Expertise is associated with deeper understanding, economical search, better memory for patterns and "chunks", automaticity of responses, better anticipation. Classic work on typists of different ages looked at reaction times and extent of preview: older typists were slower, but looked further ahead and made fewer mistakes. Expertise is domain-specific, so the issue is selective optimisation with compensation. However, experience in some domains is greater at younger ages.

Under What Conditions Are Young–Old Similarities or Differences Expected?

Age and computer experience are negatively correlated at present. An overall framework needs to recognise possible gains, to allow for younger people to be more experienced in certain cases, and to recognise that different patterns of learning can take place. Eight different conditions can be defined, varying between age-impaired, age-counteracted, age-neutral and age-enhanced.

Are There Age Differences in Learning?

This is a central issue in organisations. Without greater expertise, older workers will be less effective. Thus expertise is vital, backed by learning. However, older employees tend to do less learning, both in terms of formal training in work time, and voluntary learning in their own time. This is because of less educational experience in the past, associated with shortage of confidence, and lower motivation (self concept as non-learners). Organisational age norms focus on younger workers. This may involve, in effect, employer–employee collusion. After training, older employees are just as positive about learning: there is a virtuous cycle. Some policy makers have developed the idea of a "learning ladder", getting people onto the first steps of learning. In terms of success in learning, older people may make more errors and take longer. They may lack basic knowledge, or familiarity with terminology. They may be less expert at the outset. Literacy and learning skills may be weaker.

Are There Age Differences in Transfer of Learning?

There has been little research. Transfer involves retention and generalisation to other areas. Initial learning is vital, with over-learning and consolidation. Interference between materials in the learning environment is unhelpful. Generalisation is better when there are opportunities to use the material, individual confidence is higher and organisational support is stronger. Older people are less likely to retain new information, and less likely to generalise.

How Can Older Workers' Learning and Transfer Be Increased?

If we are successful in helping older workers, this also helps younger ones. Learning motivation needs to be increased (recognition, status, benefits). Detailed guidance is needed. Confidence needs to be increased, through pre-training and help with learning skills and strategies. We should emphasise practical activities using guided discovery. Project activities are a good basis for learning. The learning climate needs to be right: we need a combination of policy, procedures, valuing older staff, and visible support from senior management. Transfer is helped by consolidation, over-learning, refresher activities, and transfer partnership. This takes time and costs money, and the apparently selective exercise can enhance negative attitudes. Perhaps we should apply the measures to trainees of all ages, rather than singling out older workers for separate attention.

What About Non-Cognitive Features?

We should not over-emphasise the cognitive aspects of age and work, but should also consider personality, expertise and job-related motivation. Research has been limited, and we know little about age and personality at work. Are older people more conscientious, reliable and innovative? There is no solid research on age and work motivation. How are older people best motivated? We need to consider costs and benefits of work activities at different ages.

We particularly need to know more about age and job expertise, and declarative and procedural skills of workers, possibly funded by employers in specialist areas such as retail or financial services. We need to know how to create an organisational learning climate which reduces the age imbalance in learning; how personality and work motivation are linked to work performance at different ages; and whether person-person differences increase with age in personality and job motivation.

Discussion

Serge Volkoff picked up the distinction between age-variable and generation-variable factors, arguing that cross-sectional analysis can be helpful to companies. Learning problems and health problems are often linked: "experience" includes experience of oneself, and of solving personal problems. Correlation between age and experience is not always obvious, as we see in industrial processes with rare breakdowns. How can people encounter or simulate key situations? Speed is important, but complicated; there can be problems with acceleration. Complexity can be a translation issue: it has two meanings that do not correspond to young-old distinctions. Some complexity can be stupid and ill-considered, not obviously coherent, hard to deal with. Personality and intelligence issues are difficult: how do people know that they know what is needed in a given circumstance, rather than having to improvise? **Peter Warr** stress the importance of learning. Connectedness can be explored through practical activities.

Frerich Frerichs spoke about experience and innovation, and the importance of developing new products. Older research workers can remain creative, while shop-floor workers may have become preoccupied with traditional ways of doing things. This raises issues for learning organisations. **Peter Warr** noted that older people tend to describe themselves as less innovative, more conservative.

Juhani Ilmarinen asked about health and learning: how do people cope with their diseases? One way of increasing motivation is to help people cope better with illness.

Pieter Drenth noted that good and bad tended to be defined in terms of the instrumental criteria of companies. We should include issues of continuity and survival of the organisation, supervision, expressing devotion to the organisation. We should be considering non-tangibles. **Peter Warr** cited organisational citizenship behaviour studied in the USA: perhaps that should be studied in relation to age.

Geneviève Reday-Mulvey noted that older workers tended to be more interested in security, and would forego wage increases if social security (sickness benefits, pensions and training) was maintained. We need studies of older workers working part time. Evidence suggests that older workers are happy to engage in stressful work if they are able to recuperate. **Peter Warr** distinguished those who chose to work part-time, compared with those who are forced to change to part-time work.

Isto Ruoppila emphasised the importance of learning strategies: learning should not be stressful or competitive for older people, and personality factors are important. **Peter Warr** talked about stress experienced by older workers learning alongside younger colleagues: should they be handled separately? **Pieter Drenth** said that the same applies for management, reluctant to reveal their weakness, as they often know little about the technology in which they require their staff to be expert.

Roel Cremer indicated that there are a number of physical factors to consider: there should be the ingredients for a policy for learning for employers. **Per Erik Solem**

asked about compensation through use of other skills, as with typing. New abilities may reduce the need for previous ones. An overall view of different levels of Piagetian operations could show a variety of possible approaches.

Case Studies

C.H. Nygård, University of Tampere
Training for Teamwork Among Industrial Workers of Different Ages

The focus of attention was the promotion of work ability in industry. The project started based on work in the food industry, linked to new production lines 1993–94 (157 individuals). The work then moved to food, textiles, shoes and metal 1995–97 (1573 individuals). From 1998 the focus has been promotion of work ability in the separate sectors, food (600), metal (875) and textiles (1200). New work safety laws include concern for age and ageing, sex, and promotion of work ability, as well as work safety and health promotion. This provides new opportunities for research, with a growing set of respondents, including a large number of older workers. There were fewer older workers in the metal industries.

The questionnaire had been used in previous Finnish studies: concerning workers' conceptions of the enterprise, systems of remuneration, work contents, team spirit, commitment, possibilities for influence, competence, growth motivation, work ability, perceived health and physical capacity, mental stress and physical strain, sense of coherence, and sickness absence. Names could not always be used, restricting conclusions on work ability. Results were taken in 1995 and 1997, showing improvements in preconditions for team work and work ability. The shoe industry showed increases, and metal showed declines. The workforce had decreased in size over the two years. Factors such as changes in management may have been as important as the training programme that the workers followed. Team-work and work ability increased among the older workers.

M. van der Kamp
Too Old to Learn: Educational Activities of Older Workers

Max van der Kamp welcomed talk of lifelong learning, as he works in adult education.

He is concerned with the reality behind the rhetoric, as employability is ambiguous, and the tendency is to leave responsibility with the individual. Literacy and numeracy levels have been mapped by age. People need to make sense of the mass of information. Initial education is important, and work makes a difference. Considering participation levels, he used the generation model: pre-war, silent, protest, lost and new. By international standards Sweden scores well, the USA and Poland badly. When explaining low levels of participation, workers complain of a lack of knowledge; they were never informed, they thought courses were not appropriate for them. From 40 they feel too old for learning, and think in terms of retirement. Is experience an advantage or an obstacle? We think in terms of job-specific skills, work-related social network, and learning strategies. Manual workers may be more job-specific and narrow in focus, and thus vulnerable to change. Professionals think in terms of learning strategies and networking. Footballers and dancers move to new tasks within the work-related network. As we seek to make learning attractive, our concern is the transfer of knowledge and skills: workers need more abstract knowledge, not just particular packages. Meta-cognitive skills are increasingly important, but people need to be engaged in learning if they are to learn how to learn. He sought to locate adult learning in a context of abstract – concrete and formal – informal distinctions. Visits and exchanges may be seen as more appropriate and acceptable than formal education.

F. Frerichs, Institute of Gerontology, University of Dortmund
Working Conditions of Older Workers and Their Impact on
Learning and Skill Development

Frerich Frerichs described work with the European Foundation for the Improvement of Living and Working Conditions. In Germany theoretically there is no discrimination based on age, but there is data from a large study showing a decline in participation rates of older workers. The gap has widened, though participation levels for all age groups have risen. The period of expected future service is shorter for older workers. As women retire earlier, the participation in training is lower. Training programmes may not be well adjusted to the needs of older workers. Skills of older workers include familiarity with the job and the organisation, but work today does not always allow use of experience. Acquired skills may be devalued and made obsolete through new technologies. There may be intergenerational skill level differences, and age-specific skill changes. There are similar problems in service and manufacturing sectors. Women face particular problems. Difficulties are greater in Taylorist organisations, with tightly defined roles. It is necessary to create a learning environment at the workplace, develop workplace-based training schemes, combine working time arrangements and training to promote a continuous process of training, and design training courses that are effective in teaching older people.

The Institute of Gerontology identified 19 organisations who had taken appropriate measures. The chosen case study firm, Wilkhahn, in the furniture industry, enables older workers to be as productive as younger workers when introducing new forms of production. This involved a formal agreement to integrate older workers, provide special training and support training for managerial functions, targeted recruitment of older workers, and health and earnings agreements for older workers. The average age was 43.6, with 20% over 50. Training was offered to workers of all ages, adding to flexibility and adaptability of the workforce. A system of temporary job transfers was useful in broadening experience. The company needed skilled craftsmen: the approach would be harder to use in a Taylorist environment of unskilled and semi-skilled workers.

I. Ruoppila, Department of Psychology, University of Jyvaskyla
The Structure of Work Ability and its Relation to Supervisor's Age

Isto Ruoppila reported on work concerned with improving working conditions. He concentrated on the link between the work ability and ages of workers and supervisors (from shop floor supervisors to senior managers). A survey was conducted in 1987, and again in 1996. The average age of managers increased by a year, and of engineers by seven years. The working population is ageing rapidly. Educational qualifications had improved.

The Work Ability Index (WAI), developed in Helsinki, was used. It covers changes in work, unemployment, stress at work, layoffs, workload etc. In 1996 when the data was being collected, Finnish organisations were in deep recession. In 1987 unemployment was very low, while in 1996 it was four or five times higher. The pattern of employment changes after the age of 50, when some seek early retirement and others find employment hard to find. The WAI mean dips after 50, with some increase after 60 for managers. Poor work ability sets in from 50. Over the age of 55, the issues of technical competence become more pronounced.

Managers reported on changes 1991–96, listing stress factors which have increased: mental strain increased, physical strain decreased. 20% had experienced layoffs and salary cuts. In terms of salary cuts, shortened weeks and layoffs, there is little difference between age groups. Age is not a vital factor in explaining work ability. Work demands and variety of work tasks have increased: work is more active. Time pressures, mental strain and job insecurity have increased. Managers' work ability was rather good compared to other Finnish groups with similar educational background. Age was not related to work change experience

R. Cremer, NIA'TNO
The "State of the Art" on Mental Work Ability and the Increasing (Work) Life Span of People

Roel Cremer is concerned with elderly workers, and assessing ability. He is interested in both decline and growth in ability over time. He introduced two key concepts: mental workability (capacities which are wilfully allocated); and cognitive ageing (changes in mental ability, including working memory). Changes over the lifespan comprise development in breadth, accumulation of experience, and development in depth.

He looked at two categories of mental abilities, fluid and crystallised. During the career there is a shift from fluid to crystallised. In laboratory studies we find deficits, changes in performance. We can instead look from the perspective of the work environment or daily life. In the latter, there is scope for use of many strategies, thus lessening the impact of ageing. Laboratory studies focus on the abstract, ageing effects, and focus on fluid abilities. Daily life involves freer choice, where age means less, and there are fewer demands. Work involves concrete tasks, and age effects can be positive, using accumulated knowledge. Proactive policies should develop alternatives for age-sensitive professions. Researchers are concerned with determining what makes ageing workers successful, which are age friendly jobs, and what are the effects of age management at work. In conclusion, the effects of ageing are context dependent.

Roel Cremer was concerned with commitment from people, and gains experienced by the 45+ cohort. We need to look at the fuller context, including the person concerned. How do we match age-related policies with the capacities of the 45+ cohort? He presented the cyclical age-related policy used by his institute. In each cycle, researchers identify the intentions of employers, assess the present workload in terms of age-related risks and risks relating to the working conditions, and interview the employees on workability, by cohort. A report is presented, with priority recommendations, followed by corrective measures, then preventive measures for the next cohort, and finally evaluation, before the cycle is repeated.

Discussion

Lena Skiöld wondered how to persuade employers that older people are worth keeping on. The state wants people to work, but what about companies? **Pieter Drenth** defended the feasibility and profitability of key criteria other than profit figures. We need to re-educate management, taking into account the continuity and survival of the organisation. This means identification with the organisation, developing more skills than for your own job. This is what Shell call "employeeship". We need to go beyond narrow economic terms. Employees can increase output, and can contribute to the community survival. Both criteria involve hard and soft data. We need to redefine success and usefulness.

Allan Toomingas asked who is interested in the workers. Employers do not want to keep people who are below peak capacity. Is there anything unique about elderly workers? At a community level, there should be interest in keeping people at work, or the economy will collapse. We should take care of everyone. There is a need for lifelong learning. Why not have sabbaticals for older workers? Younger people would gain opportunities for experience. In his 20s and 30s he would like to reduce his working time, to enable him to spend time with his family.

Åsa Kilböm noted that women are often obliged to stop work for an extended period: they return with more energy and commitment, and with less cynicism, than male counterparts.

Juhani Ilmarinen distinguished individual, company and society levels. The key is the company level. If companies will come on board, all can be well. The priority is to develop age management. We need to change and re-train managers. Research suggests that we get more individual as we get older: we need individual solutions; how do we implement these in working life? What about the younger workers? Is it fair to bring forward individual solutions only for one age group? Each age group has strengths and weaknesses. This means problems for management.

David Wegman agreed with the focus on the employer. How do you change management perceptions? In the USA there is considerable mobility among middle managers, meaning that managers rarely face the consequences of their own decisions. There are issues for management education and development.

Max van der Kamp raised questions about small firms. They have a key role, and harder problems. **Pieter Drenth** argued that the problem might be easier in a small firm, though it is hard to send people on courses. It is important for managers to formulate longer term policies. This was, and is, the method in Shell. If senior managers do not agree, middle managers will not act constructively.

Lena Skiöld asked for evidence about contributions of older people to organisations. **Peter Warr** noted that research on work organisation and work on age have been conducted separately, and the work needs to come together. **Serge Volkoff** had partial answers. Work psychology has many tools, not always expensive, which enable us to offer interventions and suggestions at the design rather than the corrective stage. This can enable elderly workers to learn almost as easily as younger ones. It can be a matter of avoiding redundant information, ways of using symbols, in a way that enables older workers to keep up. Work organisation work has shown different approaches by old and young workers, even when both did the job well. Older workers sought, and found ways, to avoid the problems that younger workers solved. Intelligent work organisation enables both to continue. Senior managers should not take decisions which restrict the intelligent manoeuvre of more junior staff.

Pieter Drenth talked about research within the European Fifth Framework. As an aside, he noted that problems are different in the developing world of Africa, where older people are in charge, but lack insight into modern technology. This poses challenges to university authorities in areas such as information and communication technologies. Asia has a tradition of respect for the old. Those in charge can block change. One answer is to delegate to those who understand. Priority areas are Biotechnology, Health, Food, Agriculture; User Friendly Information Society; Sustainable Development and Competence Development; Energy and Environment. Cross themes are: internationalisation; implementation of research in SMEs; and development of human potential.

He offered some conclusions: ageing needs emphasis in each category; the emphasis in developing human potential tends to be on youngsters; sustainability needs the perspective of the older workforce; and older people are contributors, not just consumers.

The case was made for preparation of problem-driven research proposals. Funds have been allocated, criteria have been set, and proposals will be welcomed.

Åsa Kilböm summarised some of the conclusions from the first day of the workshop. Taking **Peter Warr's** key points, we should consider the motives there can be for employers to emphasise learning of older workers. We should clarify what we know about the context of learning in the workplace: should we mix old and young, how should we proceed? We need to spell out research needs, concerning the learning climate, young and old, learning and creativity, part-time workers, interaction between individual learning experiences and the workplace.

Alan Walker, Sheffield University
Adjusting to An Ageing Workforce in Europe – Policy and Practice

Introduction

Alan Walker considered the impact of ageing workforces on European countries, and the responses of governments. There is an increase in the average age of economically active citizens. Experience varies across Europe, comparing France with Greece, for example. Overall the workforce is ageing, but the lowering of the exit age from the workforce has meant that those over the age of 40 are seen as nearing the end of their working lives. This paradox is being addressed by the actors in the labour market. There are anomalies about early retirement "baby pensioners". In the UK some employers see competitive business reasons for employing older workers. Throughout the EU different sectors are responding to the paradox, but only a minority of employers are doing so to date.

Ageing and Employment: Early Exit

Early exit continues, with a big shake-out of older workers in EU and OECD. The position for women is less clear, given increased work involvement, but the early exit pattern seems similar. Changes are linked to demand: unemployment and recession, as well as some individual and cultural factors. Early retirement is often unemployment, as cohort analysis shows.

Changing Policy Perspectives

One reason for the pattern of change has been public policy. Older workers were seen as a solution to youth unemployment. Early exit was favoured by employers, and by trades unions. Many schemes operate at company level, with benefits seen on both sides. There was consensus in favour of early exit. Public schemes were over-subscribed. In the UK, Belgium and Netherlands schemes were popular, and based on collective bargaining. It had unforeseen consequences, was a short-term response to labour market problems, and now looks out of date. The function of public pension schemes has changed, and workers have been detached from a pension relationship, as there is a gap to be filled, in a limbo period, characterised by social exclusion and deprivation. The increase in pension costs is pushing governments to raise pension ages. The growth in early exits has devalued older workers left in the labour market, and fuelled age discrimination.

There has been a recent shift in attitudes. Governments are closing down early exits: UK, Austria, Netherlands, Sweden have all been tightening. There are twin pressures of demographic change and pension costs together with unemployment, recession and Maastricht criteria to be met. The decline in youth unemployment has allowed governments to shift the focus to other age groups. In the UK, the Welfare to Work programme has extended the age range covered.

This does not mean ageing is a priority for all nations. There are two groups: UK, France, Germany and Netherlands and others give age and employment a priority. There have been initiatives for employment and against discrimination. On the other hand in Italy and Greece it was the European Foundation research project that made age an issue. In Italy the issue has been avoided by reducing retirement age. Sweden and Finland found that high unemployment had focused attention on immigrants and the younger workers.

The engines of change are political and economic: governments want to cut public costs of pensions and welfare in Belgium France, Italy, Sweden and Finland. In Italy the focus is on falling rates of replacement. There is also the issue of extending working lives, and the informal economy, with the addition of later start to work after extended education. Germany felt prompt policies were needed to prevent skill shortages, but there are also pressures to cut pension costs. In the Netherlands and the UK there was an additional normative factor: a focus on the role of older workers, led by pressure groups seeking to reduce discrimination.

Across the European Union national political and economic pressures have dominated, some linked to the Maastricht criteria, some linked with pension costs. In 1990 the European Observatory on Age and Employment saw age and employment as a key issues, and 1993 was the Year of Older People. A survey gave a consistent set of results across Europe. Workers experience discrimination. EU documents and the European Parliament saw these issues as important, and the White Paper "Growth, Competitiveness and Employment" looked at the implications of an ageing workforce. The 1994 Essen European Council set the fight against unemployment as the central task, and required special measures on difficulties of women and older employees. In 1995 the French Presidency pressed for action on employment of older workers, for training and against exclusion: this was not mandatory. Since 1995 the focus has been on long-term and youth unemployment.

Social Partners

Employers, unions and their representatives are adjusting to new circumstances of ageing. In France and Germany the social partners often disagree, and have opposing policies on early exit. French unions wait for government to take action, and there is no perception of a need for action, so they continue to support early exit. Employers see the problems with early retirement, and look to workforce planning, skills audits, and flexibility. In Germany unions are torn between early exit and extending working life. Some unions see dangers in early exit, and want changes in the working environment to allow continuation. They argue for training to include older workers. Employers emphasise the need to cut staffing levels, and early retirement does this acceptably, allowing the promotion of junior staff. Employers support education campaigns rather than legislation. Trades unions thus have an acute dilemma. They see the injustice of discrimination, but in a context of unemployment, closures and reasonable exit packages, will often negotiate such deals. Local representatives therefore play down the negative aspects of early exit.

Good Practice in Combating Age Barriers

The age barriers project was based at the European Foundation, collecting data on good practice, examining perspectives of the social partners, and then documenting and assessing cases of good practice, and their transferability. Good practice concerned age management, creating an environment for reaching potential without disadvantage of age. Dimensions of good practice include recruitment; training, development and promotion; flexible working practice; ergonomics; attitudes in organisations; and exit policies. There are a number of impulses to develop good practice: economic and labour market (e.g. nurse shortage); changes in public policy; and organisational culture (human resource tradition, e.g. stakeholder capitalism, paternalism, need for retention of hi-tech staff).

In developing guidelines, basic lessons were sought about the key factors: backing from senior management; a supportive HR environment; commitment of ageing workers; and careful and flexible implementation. The project developed recommendations on good practice for all levels of the labour market: employers, workers, unions, national organisations, NGOs, governments, EU. A contractual obligation was suggested.

The EU may be retreating from earlier commitments, but the ageing workforce problem obliges us to address the problem seriously. Good practice benefits the organisation as a whole, and reduces the waste of human resources.

Discussion

Juhani Ilmarinen described recent developments in Finland, with three ministries concerned with ageing. A five year programme had started the previous week, with legislation including changes in work safety law and work ability promotion; training and education; health promotion and good practice. The Finnish European Presidency in 1999 will take matters further.

Åsa Kilböm reported on developments in Sweden, where recession had a paralysing effect. The situation of immigrant workers is now in focus. Public sector jobs have been reduced. Industry has been rationalising, with increased productivity. There are some new initiatives, such as Forum 50+, looking at work ability for older workers.

Lena Skiöld asked about the benefits of good practice for organisations. Alan Walker noted that an age-balanced workforce improves competitiveness. Customers like a workforce that reflects their own age. The French have found the need to retrieve collective memory, having recognised the loss of key knowledge and skills through early retirements. Companies are, above all, acting for economic advantage, and not simply in the interests of older workers.

Serge Volkoff asked about diversity within the social partners. There can be tough debate between HRM and Finance. It is a matter of support at senior level. We need to study informal discussions in companies, including the roles of researchers and consultants. This could be a useful research focus. Alan Walker noted that summary across many nations is difficult, and agreed that there is not homogeneity. He sees harder issues within trades unions, who settle for the best among available options. Employers tend to favour age discrimination legislation as a means of encouraging line managers.

Peter Warr asked again why employers bother about such issues. The number of people due to join the workforce is in decline, and skills are short. Alan Walker listed reasons for employers to bother: return on investment; preventing skill shortages; maximising recruitment potential; responding to demographic change; and promoting diversity.

Genevieve Reday-Mulvey highlighted a dilemma among employers, when rapid policy changes were made. Long-term and short-term pressures conflicted. Trades unions prefer a long-term approach. A change at senior level in companies, at national level, or at EU level, could be vital. A new general policy framework is needed. Life expectancy and good health needs to be considered.

Åsa Kilböm noted that life expectancy is going up in general, but is going down for blue collar female workers. Genevieve Reday-Mulvey argued that the number of real blue-collar jobs was falling. Åsa Kilböm looked at groups of people without income or compensation. Alan Walker noted that policy makers want a simple solution. Genevieve Reday-Mulvey emphasised the importance of the number of contribution years. Alan Walker noted that this disadvantages the long-term unemployed, where they have not paid contributions. The system needs to credit such periods.

Juhani Ilmarinen noted that the differences in life expectancy have increased in Finland. We should note the different circumstances of different groups.

Allan Toomingas considered obstacles to change: unions, employers, legislation. What about the views of individuals themselves? Do they want early exit? Alan Walker said that this is complex. From a health perspective, working conditions should be improved, not handled by facilitating early exit to an unhealthy retirement. Workers divide into two groups: some look forward to leaving with a good package, motivated by considerations of financial security. Some people drift through different difficult circumstances, often a series of redundancies, heading towards the better status of a pension.

Pascal Paoli said that to overcome the problem of health and working conditions we have to change our perspective. The ageing process itself needs to be considered.

Case Studies

G. Reday-Mulvey
Work-Time Adjustment for Older Workers in France. Lessons From a Four Year Project (1993–97)

Genevieve Reday-Mulvey, of the "Four Pillars" Work and Retirement Project, talked about adjustments in working time. She described work by the Geneva Association and for DG-V (in particular Eurowork Age '97). A recent effort has been made to redefine issues in employment for ageing workers, considering retention (including career planning, training, work design and work time adjustment) and reintegration (through self-employment and reintegration, through incentives, training and self-help organisations).

Research on work-time adjustment in 7 OECD countries was published in a book "Gradual Retirement" (1996), with support from DG-V. Management of working time means staying in working life with modified conditions. Gradual retirement is not a new phenomenon for self-employed professionals. In the past the approach has been used to reduce working life and working time, offering a transition phase, for example, in Sweden. More recently, France, Germany, Austria and Finland have passed legislation and made available incentives to reverse the trends towards early retirement, and use work-time reduction as a substitute for full early retirement. A compromise is involved between social policy (later retirement) and economic policy (earlier exit). The objective is a flexible extension of working life, increasing life expectancy in good health.

For the employer, work-time adjustment can mean reduced wage costs, increased productivity per hour, the retention of skills and expertise, reduced absenteeism, better age management, and increased job satisfaction, and it can free older workers for training duties. For employees, it can make it possible to adapt to changing abilities, reduce stress, increase job satisfaction, enhance inclusion and social integration, and increase leisure time.

On the other hand, for the employer, it raises management communication and organisation cost, and causes difficulty at first. For the employee, it can mean wage cuts, and a risk of less promotion and training.

The French scheme has operated since 1993, covering the last ten years of working life. Employers can secure subsidy only if they employ new staff from among the unemployed. Workers are in principle involved on a voluntary basis, but in practice sometimes they have been pushed. The schemes have gained in popularity, given that early retirement has become more and more impossible. At the end of 1997 over 70,000 workers were involved, without counting workers concerned with "in firm agreements" (without state help). Many workers would prefer early retirement, and would wish that earlier more generous terms were still on offer. The famous examples are from big industrial firms, but many of those involved are from the service sector, and from small firms. In some collective agreements there is training of new apprentices: this works well but only in 15% of cases, largely in the construction and manufacturing sectors. The scheme also operates for civil servants, but statistics are not available.

Two recent changes risk jeopardising the success of the scheme: the social partners pushed for early retirement in 1995 for those aged 58 and above, who had worked for 40 years, and this had bad psychological effects on other categories of workers. The financial terms have now slightly changed for firms, destabilising planning.

So what has been learned? At company level, management commitment is vital, there needs to be experience with qualified P/T work, career planning is important and social protection: must be preserved at a high level, preserving pension rights. It was proposed to establish a European network to help spread good practice. At policy level, the support of unions is needed, with a general policy framework, limits to early retirement, and limits to invalidity and unemployment, together with promotion of generous work time adjustment. This means incentives, clear stable measures, and evaluation.

Discussion

Serge Volkoff brought two examples from France. The first was an elevator factory where there had been agreement on half-time and teaching. Training is important when repairing old elevators. The scheme failed, because from the age of 52 it was traditional to transfer to white collar jobs, which workers preferred. A successful example was a glass factory, with a job sharing approach. It is a dangerous process, and information transmission was important. It succeeded because the staff were experienced shift workers, accustomed to hand-over routines. By working in pairs, they added to ownership and flexibility. The factory is in a small community, and information is shared through informal networks. **Åsa Kilböm** noted the links with debates on work organisation.

P. Drenth, Vrie Universiteit Amsterdam
NESTOR. Main Results From a Study of Ageing in the Netherlands

Pieter Drenth considered how the Netherlands approached issues of ageing as a research field. In the 1980s the work was not coordinated or systematic, and there was little interdisciplinary work in a field that needs to be multi-disciplinary. There was not enough money, so it was decided to have a centralised effort. The population was ageing rapidly at the end of the century, with 13.4% over 65 in 1996 (still lower than other EU countries). In 2020 it will be 18%, and in 2050 21.2%. Life expectancy has risen, and the increased cost of social policies is significant. The Dutch welfare state is reaching its limits, thus supporting policies of discouraging early retirement and raising pension age. Research is good but scattered.

In the late 1980s there were moves on gerontological research. A planning group on ageing decided to inventorise, set out a research programme, and develop instruments. A steering group for Research in Ageing was founded in 1982, with a research programme to be evaluated and adjusted, and making current recommendations for new 5 year programme. Efforts now focus on the Netherlands programme for research on ageing (NESTOR), whose steering committee from 1989–94 identified centres of excellence, and suggested a top-down, multidisciplinary approach to research. Research needed improved facilities and the development of support networks. Priority was given to threatened functioning, and developing possibilities for intervention, dealing with cognition and compensation in memory and attention in daily life. Clinics should address psychogeriatic disorders. Finance comes from universities, national science foundation and grants.

Results have included new collaborative work in Maastricht on memory and ageing (NMAP), looking at the impact of life events and physical fitness, cognitive ageing, memory complaints, mild head injuries, and surgical interventions. At Groningen they studied self-assessment of memory functioning, cognitive strategies, and issues of ageing when using computerised information systems. Work on economic aspects was led from Leyden, looking at exit from the labour force, post-retirement issues, and determinants of retirement behaviour. They worked with a household survey and an employers' survey, with a broad range of analysis across countries and sectors. They considered mobility, use of time, health, consumer demand, and assembled a mass of information.

The NESTOR programme had spin-off, and **Pieter Drenth** was on the review committee. Multi-disciplinary work was promoted, and new research instruments. There is now a new programme concerned with successful ageing. This includes work with new technology to optimise attention to older people. One theme is the development of adaptive training technologies for work. Does cognitive training create a general improvement, or does it only affect the specialist area concerned? A new international society has been established. The EXCELSA project involves collaborative European research.

J. Ilmarinen, Institute for Occupational Health, Finland
Maintaining Work Ability and Health Among the Middle-Aged and Elderly

Juhani Ilmarinen sketched the increase in the mean age of the Finnish population, and made projections for company cases. He took into account the risks of work disability and unemployment, showed the increasing average age, and noted the absence of young people entering the labour market. Companies with a heavy travel commitment will need to deliver with an older workforce. How will the older and younger cohorts cooperate? New government programmes started last week, which need to work at individual, enterprise and society levels. If there is failure, the bill is paid by society. Learning successful ways forward may take time, and we need good practice. In 1981 there was a longitudinal study, and were shocked by the results after 4 years, leading to a new programme from 1990–96, due for publication in English in 1998. "Finnage – respect for the ageing", includes handbooks on training, occupational health, and work ability, a total of 90 different tools. Research has considered changing attitudes of individuals and employers, and a review of legislation. Industries include metal, construction, textiles, post, firefighting, cleaning, teaching, police and vehicle inspection, as well as municipal functions.

Decreased health and functional capacity are the main reasons for early exit. This applies from the age of 55. Work stress is high, and workers do not believe employers will improve conditions. There has been an 11 year follow-up, using the Work Ability Index: work ability compared with the best, work demands, chronic diseases, handicap, sick absence, prognosis, mental ability. The Index is being used internationally. **Juhani Ilmarinen** issued an invitation to join!

Looking at groups over time, levels of work ability decline. The speed of decline is not dependent on the onset of age. Work ability declines more in physically demanding work. There are improvements and declines over eleven years, but three times as many declines. On the demand side, considering the nature of the work, decline is independent of the kind of work, mental and physical, and this applies to both men and women.

Looking at the value for prognosis, the WAI provides clear indications of those who may need future support. Sixty per cent of those measured as poor in 1981 needed a disability pension in 1992. The period from 47–51 in women is not critical, but then the increase of problems is dramatic. The same pattern is shown for men, with physical work displaying the greatest problems. In other words, the period 47–51 is the time for preventive action. The variation increases after the age of 51: this figure arises from the fact that 47 year olds were checked after 4 years.

Action is needed for those with poor work ability, looking, for example, at problems of women kitchen staff and home care workers at the age of 58. For men, those with musculo-skeletal work showed lowest work ability, and a downturn for male teachers after 51. The figures exclude those who have not left or retired. At the start there were 6500, and 5200 after eleven years. Considering teachers, in 1981 most were good or excellent, but eleven years later, there was overall decline. There is a similar story with physical installation work. We need more individual solutions, as we become more different with age. Working alone does not prevent work ability from declining. Promotion of work ability is urgently needed. Work on work ability and health was based on the follow-up data on those 500 who improved, identifying the key factors: satisfaction with supervisor's attitude, ergonomics (repetitive movements decreased) and lifestyle (physical activity increased). There is now a strong concept, based on theory and proved in practice: competence.

Starting with health, acting on all fronts from the age of 45 will improve the quality of work, and the quality of life, adding to the value of the "third age". Healthy workers are in a minority, with a dramatic change over eleven years for those of all ages. This emphasises the need for good occupational health services, requiring a great increase in expenditure. The figures may be complicated by improved diagnoses. 54% in 1992 thought that their health had declined, without diagnosed diseases. More people without diagnosed illnesses expressed confidence in their good health as they grew older.

Juhani Ilmarinen argued that physical, social and social capacities are linked, with physical changes having the greatest impact on work ability. Trunk flexion strength declines, and there are no preventive actions. Taking increased exercise made a clear difference in work ability. There is also a link with attendance at clubs or associations: if work ability is good, social activities continue, whereas if work ability is decreasing, social activity declines.

As for ergonomics, a number of jobs are physically damaging. We need: decreased repetition; better working rooms and tools; better posture; less standing at work etc.

We need to design tasks to avoid ergonomic problems. When considering recovery, we note that more is needed for heavy work, especially after the age of 40. This suggests the need for more "micro-pauses" at work. On the psycho-social side, there has been analysis of age management: attitude (to age and one's own ageing); working together (teams are better than hierarchies); organisation of work (our capacities change with age, needing individual solutions); and informing (talking, open-minded). The key is therefore age management.

Results were described from the Finnish metal industry, showing ten-fold returns on investment in health. The alternative can be disability pension before retirement age.

Results from female office workers showed improvements. If the goal is improving work ability, we have to consider capacity and health, skill and knowledge, taking the combination as professional competence, then adding the environment and individual motivation, giving an outcome in terms of work ability. This involves actions. We know what they are, but they can be hard to implement.

Discussion

Pieter Drenth asked about social capacity, and how it decreases with age. How was it operationalised? **Juhani Ilmarinen** outlined the laboratory basis for accounts of capacity. **Peter Warr** suggested that it may be a matter of activity, not capacity.

Frerich Frerichs asked why did the company concerned back this approach, and how would the message be spread to others? **Juhani Ilmarinen** noted that the key was costs of early retirement. The example is spreading.

Åsa Kilböm suggested that policy had to address all the elements leading to work ability. **Roel Cremer** asked whether employers are now more convinced of the importance of the issue. **Juhani Ilmarinen** stated that the national programme would not have been possible without this realisation. This would be a theme for the Finnish EU Presidency in 1999, with continuation under the Swedish Presidency in 2001. **David Wegman** asked about implications of a Work Ability Index score for people younger than 45. **Juhani Ilmarinen** indicated that work was being done on data covering people from the age of 29.

Case Studies

J. Nielsen, National Institute of Occupational Health, Denmark
Work Ability of Danish Employees

Jette Nielsen presented the outcomes of a Danish feasibility study. There has been a debate in Denmark, as the actual retirement age is about six years before the pension age of 67. Women and those with physically hard work leave early. Something has to be done with an ageing population. The idea is to introduce Finnish ideas to the Danish workplace, together with Work Place Health, at a company level. The Danish study is to look at work ability in a cohort of 6000 employees aged 18–65, in order to build a reference group and the capacity to identify risk groups. In addition, the plan is to test the Finnish Work Ability Index on Danish employees at a company level, and then produce a manual for Danish companies. After a couple of months there are some problems of a legal nature. Companies have been told about Finnish work, and the Health Cycle approach. Companies have to plan and run interventions, then evaluation. At least one of the firms can do all of these, and is prepared to start, with support and follow-up. The observational study comes from the Institute of Occupational Health, with plans for follow-up and consultation.

In general companies were enthusiastic, but they wanted to look at all ages, noting problems and seeking to implement any necessary preventive measures. Companies agreed to add the work to their workplace assessment. The problems come at the level of advice and consulting, where there is worry about a focus on the individual, and a missing link between the workplace and the health sector, as occupational health advice has normally been general, not individual. Finnish institutions and experience are different. Trades unions have suggested that there needs to be a policy to deal with such problems. This requires a discussion. Already there is some new legislation, but it has not yet been implemented.

Discussion

Åsa Kilböm emphasised the importance of participation. **Jette Nielsen** indicated that first the ground has to be prepared. Transfer of information about people from work to a social and health system is difficult. She welcomes reports from elsewhere.

Juhani Ilmarinen reports on over ten years of sound use of the Work Ability Index. The Finnish occupational health service is well developed. He advised that it is important to say why Work Ability is being measured. He knows of no misuse of the WAI in Finland, but there can be fears. **Jette Nielsen** described ethical, political and practical barriers at group and individual level. The objective is for companies to use the method, rather than just remaining at research level. **Pascal Paoli** described the work of a European group that looked at workplace assessment, but the key requirement is to have a policy in place at company level, setting out who will do what and to whom. It is better to have an imperfect instrument, and improve as we go along. **Serge Volkoff** described similar work in France in 1990 and 1995. The French are less worried about medical secrets, where codes apply, and there is double protection. Longitudinal work is difficult in this context.

V. Louhevaara, Finnish Institute of Occupational Health
Work Ability and Job Demands of Ageing Municipal White and Blue Collar Workers in 1981 and 1996. A Questionnaire Study

Veikko Louhevaara described a comparison between job demands and job ability in 1981 and 1996 with matched samples of workers in East Finland. In 1981 there were 50 white collar and 214 blue collar. In 1996 there were 43 and 54. In each case the age was 46–54. There have been regular questionnaire surveys. In 1996 the City of Kuopio completed a survey to promote the work ability and well being of the workers. Some questions came from 1981, and from the Work Ability Index. Job demands were thought to have increased, and physical and mental demands were too high. Subjective work ability compared to lifetime best stayed constant. There was some improvement in subjective statements of work ability related to physical demands, but no real change with respect to mental job demands. There was little change in psycho-social resources.

Overall, after two years in the present job, workers were fairly certain of their abilities. Work attitudes had improved. Observed positive differences were greater among blue collar than white collar workers. Those interviewed may have been the survivors. Ergonomic programmes had improved the workplace, with enhanced work ability as the objective, in the interim period. Each ageing worker in 1996 had participated in some of those actions.

Bart de Zwart, Coronel Institute for Occupational and Environmental Health
Successful Ageing in Physically Demanding Work. Preventive Measures in all Age Groups Required

Bart de Zwart reported on his PhD research at the University of Amsterdam. The age structure of the workforce is changing, also in e.g. the construction industry. This means increases in disability and absenteeism, linked to musculo-skeletal disorders. So, what are the links between age, physical work, and health problems? The work was based on data from questionnaires on work and health of one occupational health service. Both cross-sectional and four year follow-up analyses were done. Data was stratified by age, gender and work demands. There is a link between age and back problems, and a decline in prevalence among older workers due to early exits. Elderly nurses faced particular problems. Based on the results of the four-year follow-up study, he suggested that the problems of musculo-skeletal complaints not only starts at an older age, but also earlier. The highest risk is at age 40–49, and then health-related selection sets in.

As for policy measures, it is not enough to focus on elderly survivors only, but we have to look at the whole working career, introducing preventive measures for all workers. Demands remain constant while capacities decrease. He identified four tools. Firstly career planning: matching workload to capacity, early planning, mobility in organisation, educational system. Secondly ergonomic intervention: change tasks, results in reduction of physical load, effects for all ages. Thirdly shortening exposure: recovery problems with age, compressed workdays or workweeks. Finally physical exercise: changing maintenance capacities with age, noting that work has no training effect, and the role of fitness programmes. We need to invest in current as well as future generations of elderly workers: investing in ageing workers benefits both young and old.

Discussion

Åsa Kilböm noted that physical hard labour continues. David Wegman observed that physical workers do not then engage in physical sport. Bart de Zwart indicated that exercise is a form of recovery. Isto Ruoppila said that employer support for exercise is important. Juhani Ilmarinen agreed that recovery from hard work should happen at work. The need is for different exercise, and for a major education activity.

Åsa Kilböm asked about age and back problems, and the relation to work. She also asked about accumulated workload. Bart de Zwart said that there are problems, in that older generations were exposed to harder work in the past. Allan Toomingas asked about the follow-up study of people with back pain, and noted that the figures seemed high.

M. Millanvoye, CREAPT
Incentives to Work in an Aircraft Factory

Michel Millanvoye worked with Aerospatiale, with demand from management to find better ways of employing older workers. The proportion of older workers is increasing. He illustrated how aircraft are built, in small sections, where many jobs involve bad posture. The ergonomic study was based on interviews, then on analysis of the operator activities, in light of job requirements and operator age.

He illustrated easy and critical positions: standing was seen as good, and the exposure to impact from riveting etc. can have ageing effects on joints. Musculo-skeletal disorders are found in 72% of workers. It is similar to experience in the construction industry. It is important from the age of 40, and critical over 50. Knees, shoulders and elbows are most affected. Teams distribute tasks among themselves to avoid stress on the older staff. Easier positions were given to those of 50 and above. Younger workers face time constraints and critical positions. Ageing workers are not isolated or under-rated, and can help in training new workers.

The study covered 260 people at 50 workstations. Work was allocated according to age and strain, considered in 20–29, 30–39, 40–49 and 50–59 groups. Team working enables the roles to be distributed. It remains important to ease the critical jobs, as injuries should be reduced, and the changing age profile means that older workers will find themselves taking on these tasks, facing more exposure, and experiencing health problems. Consequently the company is hiring young people who can do the most physically demanding work. They are adjusting the workstations, and changing the assembly procedure for future aircraft.

Discussion

David Wegman was interested in the fact that assembly procedures seemed hard to change. **Michel Millanvoye** indicated that there are possibilities for change. **Pascal Paoli** said similar tasks had been faced in the automobile industry, and led to greater automation. **Michel Millanvoye** described the current limits to the use of robots. These changes had not led to staff reductions.

Serge Volkoff described the context in which the workstations have been designed, with particular levels of demand. There are links with learning, as 20–29s are pushed towards new technology, and the 30–39s miss out. Thus there is a learning problem from 30. There is not age segregation, but the figures are the outcome of statistical analysis. **Bart de Zwart** and **Alan Walker** asked how this came about. It was a matter of company policy to develop teams, and then teams informally allocate roles. **Michel Millanvoye** responded that the management were not conscious of this pattern. **Alan Walker** asked about intergenerational solidarity.

Conclusions: Research Questions

Åsa Kilböm concluded that we should pursue the full range of recommendations made by **Peter Warr**, including a number of general points about learning which require further research. It was agreed that there should be an emphasis on the role of the employer, and motives for employers to emphasise learning and competence development, and bother with older employees at all. This would include consideration of both short-term and long-term factors. There are a number of research needs in the field of learning, covering both non-cognitive features and the overall context of learning. It was clear that policies emerging from such work would have bearings on both learning and health. Bridges need to be built between policy makers and departments.

Åsa Kilböm concluded that excellent case studies are now available as a foundation point. We need to develop arguments for ageing to return to the European research agenda. In doing so we need to focus on the importance of health and work ability. This leads us to consider: the need to improve jobs; work reorganisation; ergonomics; occupational health; individual solutions; integrity and ethical issues; lifelong career development in jobs; changing attitudes about early retirement; and the mechanisms for information dissemination and networking need to be improved.

Workshop Participants

Roel Cremer, NIA'TNO, Netherlands
Pieter Drenth, Free University of Amsterdam, Netherlands
Richard Ennals, Kingston University, UK
Frerich Frerichs, Institute for Gerontology, Dortmund, Germany
Juhani Ilmarinen, Institute for Occupational Health, Finland
Max van der Kamp, University of Groningen, Netherlands
Åsa Kilböm, NIWL, Stockholm, Sweden
Veikko Louhevaara, Institute of Occupational Health, Finland
Michel Millanvoye, CREAPT, France
Jette Nielsen, National Institute of Occupational Health, Denmark
Clas-Hakan Nygard, Tampere University, Finland
Pascal Paoli, European Foundation, Dublin
Genevieve Reday-Mulvey, Geneva Association, Switzerland
Isto Ruoppila, University of Jyvaskyla, Finland
Lena Skiöld, NIWL, Sweden
Per Erik Solem, Norwegian Social Research, Norway
Alan Toomingas, NIWL, Stockholm, Sweden
Serge Volkoff, CREAPT, France
Alan Walker, Sheffield University, UK
Peter Warr, Sheffield University, UK
David Wegman, University of Massachusetts, USA
Bart de Zwart, Coronel Institute, Netherlands

Reflections on the Workshop

The workshop included superbly authoritative keynote presentations, augmented by excellent case studies and vigorous discussion. The reports on work in Finland on the Work Ability Index had particular impact, and will be taken up during the Finnish Presidency of the European Union. Issues raised in the discussions on Ageing had implications for many subsequent workshops.

2. Psychosocial Factors at Work

The conference took place in Copenhagen, 24–26 August 1998. The workshop, led by Eva Vingård and Töres Theorell, accompanied the international conference organised by ICOH, with the same title.

Abstract

During the last 25 years many things have happened in the world and on the labour market. There has been a constant global population growth, structural unemployment, an IT revolution, free global markets, gender awareness, increased education, an increasing similarity in social policies, and the continuation of wars, the threat of wars and migrating populations. This new society's characteristics will probably include a regional instead of a state-based structure and a transformation from a local to a global perspective. Transnational companies will create large centres of creativity and small units of production and distribution near the market. There will be an increase of mobility of people and information, and increasing complexity. In future the labour force is likely to be divided into 20% attractive elite workers, 40% highly skilled, 20% low paid flexible workers and 20% unemployed.

This scenario will have a strong impact on psychological and social factors for individuals, groups of employees, families and societies. To measure the health consequences, different models have been proposed. One example is the demand-control model (demands from the job situation related to the degree of control) by Karasek and Theorell, supplemented with the degree of social support from supervisors and work mates, introduced by Jeff Johnson. This model provides the basis for many redesign efforts aimed at improving the work environment.

Another model of interest is the effort/reward model proposed by Johannes Siegrist, concerned with the degree of effort and the amount of reward achieved from the job. An imbalance, with high effort and low reward, is associated with adverse health effects. Coping strategies are of interest, as well as the balance between skill and demand. In recent years empowerment as a basis for redesign has been discussed as a health factor. Negative life events, and their relationships with different health outcomes such as ischaemic heart disease and psychiatric disorders among women, have been shown.

To measure the relevant psychological and social factors, different questionnaires and interview techniques have been developed. The intrinsic and extrinsic validity of these instruments in different societies needs to be discussed, together with considerations of the ethics and morality of using them.

General Reflections on the Conference

There can be no single report on the dialogue in the parallel sessions of an international conference. As a newcomer to the specialist research community, I formulated some questions before arriving:

1. *Is there an apparent Scandinavian or Northern European agenda which can form part of the Swedish Work Life 2000 programme for Europe, to be launched at Malmö in January 2001?*

2. *What are the presuppositions on which that agenda rests? Are they shared across Europe? What are the obstacles to progress?*

3. *Is it assumed that there is a central role for government researchers and research institutes? How does that fit in with an increased role for market forces and a reduced role for the state?*

4. *There are costs associated with addressing the psychosocial factors at work. Who is to pay?*

5. *Should psychosocial factors be best considered in the context of organisational development and change, or in isolation in "purely scientific" terms?*

6. *Should we be proposing new international collaborative projects, to report at Malmö 2001, but also linking to earlier EU Presidency agendas? Should we then involve partners beyond Europe?*

A reading of the papers suggested considerable diversity, with a core of session leaders and some leading projects, the subjects of several sessions. There is an established research community with international networking, resulting in some joint papers from more than one country, including valuable comparative studies. The range of topics covered extended across much of the agenda of Work Life 2000.

The research implicitly depends on favourable or tolerant political environments, as well as financial support. Presentations appeared to assume an ongoing calm medium term economic situation, allowing the continuation of conventional research. The British group at the conference, led by **Tom Cox** *of Nottingham University, was small, and represented a community with limited funding but a long tradition.*

There was discussion of intervention, which is presented as separate from, but could be construed as being aligned with, management. In a number of European cases it was explicitly assumed that the workforce would be able to choose between forms of work organisation, including by forms of voting. The US research tradition had focused on the individual, while the Scandinavian tradition had been organisational.

The conference organisation and environment placed some distance between presentations and the workplace. Many accounts came from researchers outside the work practice, constituting predominantly quantitative studies of issues whose qualitative nature is poorly defined. There were exceptions to this, with accounts from the Swedish MOA programme based on analysis of interview transcripts, which succeeded in bringing the session to life. In a number of cases the researcher appeared to think that the task was complete when a descriptive account of the workplace had been given, and that the development of policies and strategies to address the situation was a matter for others. This illustrates the problem of research dissemination, the subject of a separate workshop.

First Day of Dialogue

Tage S. Kristensen raised challenging questions about intervention, arguing that researchers and their funding agencies can be selective in their procedures, often neglecting randomised controlled trials, and disregarding negative and inconvenient findings. He argued that qualitative approaches are appropriate during the process of research, with quantitative approaches applied at the endpoint. Many studies have suffered from a lack of control and follow-up, have lacked generalisability and comparability, and seem devoid of theory. In a rapid overview of past research internationally, he gave recommendations for future approaches.

On the first day, vigorous discussions followed the presentations by **Eva Bejerot**, **Sarah Thomsen**, **Per Wiklund** and **Annika Haremstam**.

Eva Bejerot's account of change and its impacts on those concerned with "life" and "things" can be seen together with **Sarah Thomsen's** comparison of mental health nurses in Stockholm and Birmingham (an earlier study compared psychiatrists). Management models which were perhaps more appropriate to "things" were applied to professionals concerned with "life". In the UK hospitals have been run as businesses, concerned with profitability and cost reduction, and with medical judgements no longer given primacy. Accounts of the management dimension were given in the symposium on downsizing and re-engineering, and in particular in the paper by **Aslaug Mikkelsen**.

Many papers presented conventional quantitative analyses of questionnaire responses, interpreted in terms of psychosocial factors. Two presentations describing work of the same team took a different approach. **Per Wiklund** undertook detailed analysis of interview transcripts, deriving common script structures, and isolating key phrases and sentences which highlighted the feelings. The session participants agreed that the accounts resonated with their experiences of hospitals as places with underlying tensions and battles. He was unclear what will happen next, and has yet to give full feedback to those he studied. For many occupational psychologists, the move from description to active intervention is difficult or impossible.

Annika Harenstam of the MOA project was in similar territory, with her exposition of the views of women in the struggle for human dignity and a position in working life, based on interviews and a developing model to explain what is going on. If we see it as inappropriate to apply the market model to human services, and we note that accounts of work tend to exclude consideration of unpaid domestic activity, then we must have doubts about the wisdom of straightforward reliance on quantitative data derived from questionnaires framed within such flawed assumptions. The qualitative dimension is vital, and it can reawaken appeals to our common humanity, enabling us to consider relationships and actions outside the business metaphor.

There is a rich tradition of Swedish work on which to build. The work of **Ingela Josefson** with doctors and nurses, conducted at the Swedish National Institute for Working Life, appears relevant to the MOA study. Her 1988 article on "Nurses as Knowledge Engineers" and her subsequent work using Greek tragedies with doctors on professional development programmes, are rich sources of insight. She encourages professional reflection on working life, and the recognition that works of literature may deal with relevant themes. In the first case the external stimulus was information technology and artificial intelligence, in the second the cathartic potential of Sophocles. Medicine and the associated dilemmas and problems are as old as civilisation.

The Dialogue Seminar Workshop, initiated by **Bo Goranzon**, now at the School of Industrial Management at the Royal Institute of Technology in Stockholm, provides another potential means of taking working life to the stage and a wider audience. He has led work on "Philosophical Dialogues", a kind of intellectual forum theatre. His actor and director colleagues have the skill to take an interview transcript or dialogue, and give it dramatic life.

Second Day of Dialogue

Norito Kawakami gave an excellent and provocative presentation on "Psychosocial working conditions and mental health", demonstrating that many of the supposed differences between Japanese and US workers, when seeking to correlate work over-load and depression, for example, can be explained in cultural terms. He outlined different models of self in East and West, with the Eastern tradition being one of relating to others and fitting in, and the Western tradition of independence and standing out. There are also differences in patterns of responding to positive and negative wording in questionnaires, which can provide major distortions if not rectified through revised wording. His presentation cited core research results from European ICOH members, and he was able to expose the weakness of quantitative analysis that neglects the cultural dimension.

Gunnel Ahlberg-Hulton discussed changes in the work environment of healthcare workers. The paper was based on a study in the Swedish community of Norrtalje, which has undergone particular patterns of social and economic change, including increased unemployment, which were not given attention in the presentation.

Aleck Samuel Ostry's presentation on sawmills in British Columbia followed, with an account of major industrial reorganisation and change between 1965 and 1997. These changes parallel those in the Norrtalje region. **Ostry** described a small group who have survived changes which have had a traumatic effect on the community. Discussion after the presentation highlighted parallels with the experience of Malmö.

Per Wiklund then discussed a qualitative account of boundaryless work. He is a rich source of material which is capable of wider presentation. His account of flexibility and change in the work process is intelligent and clear, explaining the shift to trans-formation rather than production. He moves into the territory of psychological problems of individual feelings of abandonment, and socialisation into working life. He described what he called a "Catch-22" solution as the change is attempted, outlined the pressures of the demand for high performance, and presented the indi-vidual as a sign. Individual conflicts are seen as following the weakening of regula-tions. **Per Wiklund** was asked about consequences for job development in light of his account. He responded that workers need help in establishing their own bound-aries. He could have referred to work in, for example, learning regions, as supported by NIWL.

Kurt Baneryd outlined Swedish supervision programmes for stress reduction. Government objectives were benign. He distinguished the experience of men and women, but, for both genders, teachers and doctors faced the greatest stress and lack social support, with low control experienced by workers in pulp and paper production.

The poster session demonstrated particular active interest and involvement from Brazil. There are major practical problems to be overcome before stable research can be undertaken, but it is hoped that the next conference in the series can be held in Rio de Janeiro. There is evidence of interest across the industrial sectors, and the Brazilian economy has significant involvement from multinational corporations.

Third Day of Dialogue

The third day constituted a symposium open to a wider audience, and attracting 100 Danish occupational health practitioners.

Töres Theorell gave a fluent overview of the future of healthy work, drawing on rich Swedish data. He located the discussion in a global context, comparing American and Swedish perspectives. Women are taking on more demanding roles in physical and psychological terms, and the problem of awkward postures has increased. Life experiences of the well educated are different from those of the poorly educated. He discussed the demand–control–support model, and data from surveys to date, and extrapolated. Increased pressure at work is producing exhaustion and burnout. There are clearly documented health effects of long working hours. How hard are we going to push people? We know that normal behaviour and arousal patterns are being disrupted. Clinical depression can be seen as an outcome of "brake failure". Chronic fatigue syndrome is "accelerator failure". Post Traumatic Stress Disorder can arise from the workplace. There are three separate disorders, requiring separate treatments. He argued for physiological descriptions, rather than "burnout", derived from responses to questionnaires.

There is debate about physical as opposed to psychosocial data, while the brain makes no distinction. The MUSIC-Norrtalje study clearly confirms this. When one system malfunctions, other systems may react more vigorously. We do not understand the causes of chronic fatigue syndrome. We must expect new chemical circumstances, vibration, radiation and other physical problems at work as well as psychosocial factors. We have probably overestimated the significance of cognitive functions.

Unemployment and pressure for effectiveness will reduce active comment and complaint from workers: there is a growing silence. Here is a case for international collaboration.

There has been research on unemployed people working in the regions on meaningful activities. Positive mass media reporting can distort individual responses. Physiologically there can be deterioration. We should be using physiology more in this field. He cited his own research on the consequences of support from the employer. He highlighted gender differences in experiencing working conditions. Coping when facing unfair treatment at work varies by age and gender. This links with blood pressure and hypertension.

He welcomed discussion of intervention, and the translation of theory into practice. There are questions regarding sample sizes. Intervention should not be rigid and impractical, but pragmatic. It is important to know how to do the intervention before starting the evaluation. In conclusion, physiological factors will increase in importance, in practice and in research. Neurobiology will help our understanding and monitoring of working life, when we will face denial and silence. There needs to be funding for research on weaker groups, who should not be left to deal with their problems alone.

Robert Karasek considered the global dimension of the political economy of psychosocial job design: tabulating costs, training new behaviours, developing a new overview language, developing social support and speaking out multi-sided economic development. *Fortune* magazine has declared that "the job is over". *Business Week* considered the rewriting of the social contract. There are models for change, with implications for social policy, for example, deregulation. They are presented as simply logical, but there are alternatives. "Leaner, meaner, and going nowhere faster" reports the *New York Times*. *Business Week* describes "the horizontal corporation" where employees can be trusted to do the right thing. Psychosocial work design was dealing with this decades ago. *Fortune* now talks about "intellectual capital". The methods appear to have been used to reach the wrong goals. New goals are needed to replace profit maximisation. The goals of conference participants are a key resource.

Robert Karasek described visits to different countries where cuts in social costs have been recommended as a means of competing globally. Both competitors cut welfare costs, and continue to compete and cut. Economists are happy, but psychosocial researchers know the costs of these changes, which have not been considered. He described an international Job Content Questionnaire in the Netherlands, Japan, Canada and the USA. How could scales be developed for cross-cultural analysis? They would need to take account of skill discretion, decision authority and psychological demand. Across the nations the standard deviations and correlations are similar, with the greatest differences appearing in psychological demands. He noted problems in occupational coding systems. There are differences in rankings, with more consistency within occupations than countries. The global economy appears to have arrived.

He then discussed interventions, and pathways to a humane sustainable social future. This meant a dynamic active demand–control–support model. We need to identify new kinds of productivity, recognising that the workplace is part of democracy. The answer is to be found in the active dimension. Michael Porter's model, as revised by **Robert Karasek**, considers the basis of modern economic activity. The four factors of production are seen as: human resources; active smart customers; motivation in production; and supplier networks.

Robert Karasek introduced the idea of conducive production: helping development of new skills and capabilities. The key products are tools, promoting capabilities. For example, the word processor enables new levels of activity, requiring new tools. There is demand for intelligent contributions, not stupidified work. This relates to the demand–control–support model of psychosocial factors. He referred to a study in a Swedish nursing home, getting to know the patients better (and leading to the patients wanting to take over management of the home). Absenteeism rates among staff changed, reflecting the proper use of staff skills. This illustrates productivity as increased capability.

New tools are needed for new productivity. Taylorism had called for compartmentalised understanding. Conducive production requires coordination, horizontal communication, and new tools. Value is derived from association. He introduced the "skills plate", which was used in south-west Sweden, studying an engineering plant and the use of skills. Many available skills were not being used. Skills plates can be assembled to form a picture of skills sets and links. Problems can be analysed and solved. The workers had not previously been asked to participate and take an overview. The discussion was launched, and researchers were no longer needed.

Robert Karasek then reflected on the Internet, not developed for short-term profit but to link capabilities of researchers. Connectivity of capabilities was the essence. Java offers portability across platforms, but Microsoft have tried to re-tailor it to Windows. This constitutes a conflict of models, a threat to the development of capability.

The Conducive Value Economy has to build on the Commodity Value Economy. Alongside we can add new ideas, subsidised from profits. The two can link. Robert Putnam has written about the civic society and social trust. The key is to build trust.

The conducive economy is how it was before. This highlights the danger of social vacuum. **Robert Karasek** cited Wilhelm Reich's view of Germany in the 1930s, and the absence of social structure. He feared for the future. Psychosocial researchers have a contribution to make, helping redefine the economy and society. This changes the social dialogue. He ended by noting that the audience understands much about productivity and the costs of economic development. Occupational health is not just about treating the wounded, but engaging in active change.

In discussion it was noted that hindering productivity leads to illness; thus competitiveness, productivity, health and effectiveness are linked. The model requires the researcher to be active, to be concerned with change, to address the organisation as well as the individual.

Peter Hasle is research director of the Centre for Alternative Social Analysis, in Copenhagen, a privately run not-for-profit consultancy, involved in the work to reduce monotonous and repetitive work in Denmark, with a target reduction of 50% 1993-2000. This followed a campaign by textile workers. The plan is delivered through a coordinating committee and industrial working environment councils, through the social partners. Companies make action plans, and there is support from the labour inspectorate, with government funds for training, infrastructure and support. Repetitive work is defined as lasting for more than 3–4 hours daily, characterised by work cycles of less than 30 seconds, or the same movement for more than half the cycle time. This is defined as hazardous.

A mid-term evaluation was conducted in 1998, starting with RSI-intensive industries. There have been visits, telephone interviews, and targeted interviews with key players. There are clearly methodological problems: there is no clear baseline for the prevalence of this work in 1993, it is hard to remember past cases, the sample has been limited, and health hazards have not been studied. It is still worthwhile to undertake the evaluation, as early results are needed for practical and political purposes. There are positive results, with awareness of the action plan and programmes leading to a 25% reduction in the first part of the period. There have been secondary benefits, including productivity improvements. However, most companies had not made proper assessments or plans. Easy solutions have been used, and the success has often been overestimated. We cannot expect further major reductions. Job rotation has been seen as the solution. Overall, the effect has been successful, thanks to cooperation by employers, unions and labour inspectors. Long-term thinking is necessary, combating deep-seated Taylorism.

The case study demonstrates the Scandinavian approach, relying on almost complete unionisation. Occupational health services have been important partners. Safety organisations have a more short-term perspective. The research did not suggest that the problem has been solved by exporting jobs to low wage countries. There were cases where approaches other than job rotation proved effective: in the metal industries the removal of monotony has formed part of job redesign, and has been seen as part of Human Resource Development. The study took place at a time and in sectors of low unemployment. To date there has been no penalty for non-compliance, but this issue will arise in 2000.

Michiel Kompier from the Netherlands spoke about stress prevention, using a participative stepwise approach. There have been previous reviews of work on stress prevention. There has been considerable activity, in which the focus is on reducing effects, after the event. The main target is the individual, not the organisation. There is a lack of serious research into effects. Systematic risk assessment is absent. Why is this the case? Individuals have been seen as reflecting opinions of company management. Psychology has been based on individual cases. Intervention has presented methodological difficulties. Costs and benefits of prevention have been neglected.

What can we learn from the review? For too long prevention has been based on charity or the requirements of law. We need to address the hard outcomes, and present examples of good preventive practice, taking advantage of a multi-disciplinary approach.

He went on to evaluate ten "natural experiments" in the Netherlands. He wanted primary preventive measures, stress audit, support from companies, and the agreement of minimum methodological standards. Breaking the process into five steps, he compared the cases, which included a government department, a prison, a cigarette manufacturer, an oil refinery, a building company, a hospital, a home for the elderly and a telecoms company. Companies were motivated to participate due to concerns over absenteeism, psychosocial and musculo-skeletal load, high costs and inefficiency, and the decreasingly social nature of work. Staff turnover caused unwelcome expense. Projects included representation from the parties concerned, with an active role for employees, and external support. After problem analysis measures chosen included job redesign, sickness absenteeism management, and initiatives in Human Resource Management and training. Bureaucracy could be seen as obstructive, and the coordination of the programme posed challenges, some of which were overcome through experience. It was important that the process was stepwise and systematic, involving diagnosis and risk analysis. It was both work-directed and worker-directed, it was participative, and enjoyed management support.

In conclusion, the benefits of prevention should be stress. Programmes work best if they combine bottom-up and top-down approaches, and if they combine natural experiments with cross-sectional studies. In discussion, reliance on self-reporting was challenged. However, among the objectives was the search for good practice; thus a degree of self-selection was acceptable.

Aslaug Mikkelsen described a participatory approach to organisational intervention on work environment and stress in Norway. Health and Working Life is a national programme. There were four kinds of intervention, carefully timed and coordinated. Organisational interventions came in Post Offices, following interviews. This built on traditions of participation, dialogue and democracy. Reference was made to the literature, with an emphasis on organisational learning. Search conference methods were used, generating proposals. The research coincided with major restructuring, which had a major impact on the case study workplaces. Future work needs to develop organisational interventions and practical case studies.

Wilmar Schaufeli from Utrecht gave an account of burnout in human service work, starting with three informal personal case studies. He spoke as a researcher and clinician. A recent EU study showed major effects of work on health. Mental disorders have risen fastest. Each day 160 new mental disorder diagnoses are made in the Netherlands. Burnout has not been recognised: it is a new phenomenon, linked to the growth of the service sector. Labels have changed, there is a greatly increased emotional workload, weakening of authority, yet continually rising expectations. This often constitutes an effective breaking of the psychological contract. Swedish researchers have challenged the definition of burnout as a distinct syndrome. He argued that the key distinctive factor is depersonalisation following exhaustion. It is prevalent among teachers and those working in social services and medicine, particularly mental health. **Wilmar Schaufeli** concluded with an account of burnout workshops, and their effectiveness.

Workshop Participants

Gunnel Ahlberg-Hulton, Karolinska Hospital, Sweden
Kurt Baneryd, Swedish National Board of Occupational Safety and Health, Sweden
Eva Bejerot, NIWL, Sweden
Tom Cox, Nottingham University, UK
Richard Ennals, Kingston University, UK
Annika Haremstam, Karolinska Hospital, Sweden
Peter Hasle, Centre for Alternative Social Analysis, Denmark
Robert Karasek, University of Massachusetts, USA
Norito Kawakami, Gifu University School of Medicine, Japan
Michel Kompier, University of Nijmegen, Netherlands
Tage S. Kristensen, National Institute of Occupational Health, Denmark
Aslaug Mikkelsen, Glostrup Hospital, Denmark
Aleck Samuel Ostry, University of British Columbia, Canada
Wilmar Schaufeli, Utrecht University, Netherlands
Lena Skiöld, NIWL, Sweden
Sarah Thomsen, National Institute for Psychosocial Factors and Health, Sweden
Töres Theorell, National Institute for Psychosocial Factors and Health, Sweden
Eva Vingård, NIWL, Sweden
Per Wiklund, Karolinska Hospital, Sweden

Reflections on the Workshop

The set of Swedish research presentations at the conference embody implicit assumptions about the roles of researchers and governments, including relations with public and private sector organisations. There is a characteristic focus on delivering results by traditional methodologies, and a reluctance by some established researchers to consider alternative approaches, and to find common ground with researchers in other fields. The move from research to action through action research has been made by some, while others need help to follow. Many Swedish researchers will have been heartened to find some of their long-held positions endorsed by international colleagues.

My approach in parallel sessions and symposia was to ask questions and suggest connections, testing the perceived boundaries between research fields and sub-fields. This also served to clarify the point that researchers have attitudes related to the institutional cultures where they work, which in some cases correspond with those of national government sponsors. There is an underpinning layer of institutional attitudes which may be exposed in discussion at a scientific conference, and need to be understood when planning a process and conference such as Work Life 2000.

There are some interesting issues of definition to be addressed. The phrase "Work Organisation" has suddenly achieved prominence in 1998, but with a multiplicity of different definitions and agendas.

For example, the UK European Presidency conference on Work Organisation in Glasgow, held in April 1998, included no mention of occupational health, although this would have pleased DG-V. The UK starting point had been an emphasis on the employability and flexibility of the individual, rather than the adaptability of the organisation, and there has been a reluctance to discuss regulation. However, British adherence to the European Social Chapter and the introduction of a new White Paper "Fairness at Work", within a framework of employment guidelines covering all member states, suggests that definitions and attitudes are shifting.

By contrast, some have regarded the ICOH conference in Copenhagen on "Psychosocial Factors at Work" as a conference on Work Organisation. One intriguing aspect of this convergence of interests, without yet sharing definitions, is that there are numerous cases emerging of researchers in the same universities, with shared interests, who have never met or even heard of each other. We are witnessing "Partnership for a New Organisation of Academic Work". One of the functions of Work Life 2000, with a wider all-encompassing brief within which Work Organisation, however defined, is one of five themes, is to be a midwife to the academic world of the next Millennium.

3. Managing and Accounting for Human Capital

The Workshop was led by Jan-Erik Gröjer and Ulf Johanson of Stockholm University, and held at the Office of the Swedish Trade Unions in Brussels, 14–15 September 1998.

Abstract

The managing of intangible assets is a key issue in the creation of wealth for individuals, organisations and nations. In the past the European Commission has discussed appropriate measures to increase the transparency of intangibles, especially those concerned with human capital. However, too little is known about how the intangibles are managed and accounted for, and how they contribute to growth and employment.

The workshop leaders are responsible for two workshops, the first on Human Capital, and a further workshop in 1999 on Intangibles. Their intention is to produce a paper setting out policy recommendations. The aim of the first workshop was to formulate normative statements based on the state of the art with respect to human resource costing and accounting in the following areas:

1. *Human capital information* in the annual statement and in relation to efficient capital markets. The underlying assumption is that agents on the capital market (or even other stakeholders) cannot interpret and value the modern human resource organisation in an efficient way.

2. *Human capital investment models* and the capital expenditure problem (from an individual, an organisational and a societal point of view). The underlying assumption is that organisations underinvest in human resources due to the lack of appropriate investment models and/or human resource information.

Special emphasis was to be devoted to competence development. Additionally, the workshop was to consider the human capital environment, especially the emerging new forms of employment conditions.

The first day of dialogue was spent exploring the conceptual framework of the subject, involving all participants. The agenda was present as a target objective, rather than as the basis for discussion. As an initial guide, participants were referred to the following outline:

Human Resource Costing and Accounting – Its Definition

- Are we trying to represent human beings, employees or human artefacts?

THE STATE OF THE ART

The Need for Representations (the Demand Function)

- From a return on investment point of view?

- From a management control point of view?

- From a capital market/corporate governance point of view?

- From a societal (macro) point of view?

This point on the agenda has to do with the evidence we have of the underlying problems and opportunities of Human Resource Capital Accounting.

Existing Solutions (the Supply Function)

- How do organisations and society cope with HRCA – empirical evidence?

- Suggested theories and models fruitful for future application/use?

POLICY IMPLICATIONS

- The need for accounting or other regulation in an EU perspective

- The need for model development and testing

- The need for future research and research organisation

Reflections on Human Capital: Characters in Search of an Agenda

The workshop began with a free-ranging discussion of the issues, without formal presentations or papers. **Ulf Johanson** circulated a comprehensive literature survey, undertaken by Stockholm University for OECD. **Alfred Gutschelhofer** from Austria presented conclusions from a recent survey of accounting academics. **Robin Roslender** presented a picture of conservatism among accountants and accountancy researchers in the UK. **Per Bukh** reported on Danish approaches to Intellectual Capital Accounts. **Guy Ahonen** outlined the situation in Finland. **Thomas Gunther** launched a search for a marketing strategy for the new topic. **Laurie Bassi** described findings from the American Society for Training and Development, with additional insights from **Gregory Wurzburg** of OECD.

Conventional scientific reporting encounters problems when the nature of the science in question is being challenged and the alternatives are contested. The old orthodoxy has crumbled, but there are competing blueprints for the new. The example of environmental accounting was quoted to demonstrate that change is possible; **Robin Roslender** noted that new narratives are beginning to appear, but he wondered who was to do the reporting in the age of outsourcing and the virtual organisation.

Work Life 2000 and Human Capital

The focus on "working life" rather than on "business" implies a different, but unspecified, way of valuing, managing and accounting for human capital. As **Robin Roslender** argued, there is a comforting appeal in the prospect of an accepted supporting methodology, making work with people as acceptable in accounting terms as work with money or things. As many participants noted, it is apparent that

the new concern for "knowledge management" reflects a realisation of the importance of knowledge, and human capital, as opposed to traditional physical assets.
However, we may feel that it is often an enthusiasm expressed by those who themselves have little knowledge, but note that the value of their company as perceived by
the market exceeds the value of other assets, and should be explained (accounted
for) and managed in some way. **Laurie Bassi** outlined work with seven major multinational corporations, seeking to find a uniform way of describing this situation.
This begins to bring together ideas of accounting and accountability.

Human Resource Development, to be a useful approach, needs to reflect changes in
work organisation, and the changed nature of working life, career development and
working patterns. **Gregory Wurzburg** noted that change often meant that external
factors were converted to internal costs, and people could be seen as liabilities. **Jittie
Brandsma** argued that investing in people has implications beyond the workplace,
for education, training and community development. For companies to sustain their
development over time, they need to network with stakeholders. Governments have
frequently seen training as a way of reducing unemployment figures, and the success
of schemes has been measured in those terms, rather than by evaluating content.
Companies have been reluctant to train individuals, as this increases the likelihood
of their being "poached" by rival employers. In the UK, Investors in People is a
means of encouraging Human Resource Development, offering marketing benefits
from accreditation.

Guy Ahonen set out the agenda from Finland, where industry had taken the lead,
needing to find justifications for high labour costs in a global competitive market.
The critical need was to innovate. Nokia was taken as the motivating case study.
Related experience in the Swedish Working Life Fund, with 25,000 workplace
projects, which has been reported by Bjorn Gustavsen, was that expenditure on
training showed little or no correlation with increased productivity, unless this was
combined with programmes of organisational development, which led to major
improvements. These conclusions are now being taken seriously by other governments, including the UK. In the United States, companies need help in assessing the
return on their training investment, or in presenting incontrovertible non-financial
arguments for expenditure in an area where the details have not normally been
reported. **Alfred Gutschelhofer** argued the case for key indicators, which **Laurie
Bassi** and **Ulf Johanson** agreed should be as simple as possible.

Ulf Johanson introduced the literature review, noting that the papers cited had been
located in different academic disciplines, with very few addressing ethical issues. He
raised the question whether information, once disclosed, was always used internally.
Research was lacking, and typically based on small numbers of interviews with
CEOs. **Laurie Bassi** cited studies by Ernst & Young on related leadership issues. She
emphasised the importance large companies placed on obtaining proper information to support investor and analyst confidence, while the small companies quoted
by **Per Bukh** in Denmark were largely unquoted, and seeking to develop their future
market potential. **Gregory Wurzburg** summarised research from the Brookings
Institute on intangibles, and highlighted conflicts of interest faced by analysts.
Jan-Erik Gröjer offered a conspiracy theory for senior management reluctance to
incorporate new kinds of information, as this would force changes in company
management, such as the inclusion of Human Resources directors at board level.
More generally, **Ulf Johanson** argued that those with power do not wish to change.
Jan-Erik Gröjer noted that Daimler-Benz seemed to use foreign managers as a

means of effecting change, but observed that across major companies there is a tendency for Chief Financial Officers to become Chief Executive Officers.

There was an unsuccessful attempt to agree on a new phrase to capture the successor to conventional accounting, applicable internationally. **Robin Roslender** spoke of "Human Worth Accounting", and expounded a societal point of view, to which **Jan-Erik Gröjer** responded that there is no such thing as society in general, a lesson, he argued, that had been learned from "Social Accounting". **Robin Roslender** wanted to demonstrate if such an approach is possible, while **Laurie Bassi** wanted to see if it was correct. **Guy Ahonen** had grave doubts as to the advisability of placing people on the balance sheet: his search was for appropriate structures and indicators.

Broadening the scope, **Jan-Erik Gröjer** spoke of "Value Contribution Accounting" and "Knowledge Management". **Jittie Brandsma** discussed "Learning Organisations", and how to become one. **Ulf Johanson** asked about "Intellectual Capital Accounting", and **Robin Roslender** responded with "Human Competence Accounting". **Alfred Gutschelhofer** prefers "Human Asset Accounting", with a central role for information systems, and a need for multidisciplinary approaches. There was considerable discussion of the "Balanced Score Card" approach.

At this stage, **Guy Ahonen** showed a diagram from his recent best-selling book, in which he set out the "Real Balance Sheet", with two columns headers Assets and Capital. He offered the insight that it was a balance in the Invisibles section in the Capital column that led to the pressure to define the corresponding Intangibles in the Assets column. He breaks Intangibles down into External Structure, Internal Structure and Competence (individual and group). **Jittie Brandsma** indicated that there are major differences between countries and languages in the ways in which concepts such as competence and qualifications can be discussed.

The alternatives continued, with **Ulf Johanson** arguing the case for "Human Resource Accounting". **Jan-Erik Gröjer** saw the need for a new book. **Gregory Wurzburg** thought that many of the suggestions showed the messiness of mixing disciplines and languages, although "Intellectual Capital" captures an important idea. **Jan-Erik Gröjer** argued that the market will determine which concept survives. The flow continued unabated: "Human Value Management", "Human Worth Accounting", and many more.

Whatever the concept is called, the need is there. **Stefano Zambon** argued that we need to separate consideration of management accounting and external reporting. **Jan-Erik Gröjer** noted that the concern in Sweden arose from the need to improve investment in personnel, so the basic motive was to improve business efficiency. **Laurie Bassi** reported that in the USA interest is in maximising return on investment, and that this could be achieved in part using non-financial measures of value. In response to **Ulf Johanson**, **Guy Ahonen**, **Laurie Bassi** and **Per Bukh** agreed on the important role of equity markets and the need to develop a small set of indicators. **Jan-Erik Gröjer** and **Laurie Bassi** discussed risk awareness and insurance, and agreed that to some extent doing what others are doing is a form of insurance. **Laurie Bassi** cited the view "If you can't measure it, you don't need to do it".

Are the concepts complete and ready to be marketed? **Guy Ahonen** pointed to gaps, such as knowledge inventories. **Laurie Bassi** described ongoing research on investment in people and financial benefits, with correlations that suggest causal links. **Guy Ahonen** responded that if we have HR indicators, then we can see improvements as based on HR investments, otherwise we simply don't know where the money has come from.

Ulf Johanson raised questions of power. **Laurie Bassi** suggested that an absence of accountability could be fun, if you can get away with it. **Gregory Wurzburg** pointed to experimental work at Motorola on training and cost minimisation, which **Laurie Bassi** argued had to be seen in the context of the Motorola commitment to education.

Jan-Erik Gröjer wanted to consider policy recommendations, even though there was not full agreement on concepts. There is a need for more research, with a particular focus, suggested **Jittie Brandsma**, on returns on investment in training, where we know remarkably little. **Gregory Wurzburg** asked for a clearer delineation of the area of working life to be covered. **Jan-Erik Gröjer** responded that in a context of increasing competitiveness, insights into working life are an important Swedish contribution. **Laurie Bassi** noted that this did not mean that Europe was seeking to follow the US model, and recalled that the conclusion of the June EU-US Workshop on Work Organisation was that a combination of efficiency and equality, security and flexibility was required.

Jan-Erik Gröjer, over lunch, had put forward an informal case for a new Directive or international convention banning Human Resource Costing and Accounting, ostensibly on the basis that it gave unfair competitive advantage, but with the consequence of tacitly encouraging the practice. It is clear that financial analysts consider such issues when forming their views on particular companies, but without presenting details in their reports: this information is seen as privileged and valuable. Indeed, compulsory compilation of such information would pose a threat to the privileged employment position of analysts. Yet again we have evidence to support the European view that there is no one best way, that we have much to learn from differences, and that creative progress can be made through networks and coalitions.

Policy Implications

Ulf Johanson set the agenda for the second day, based on the three questions in the original background paper, concerning Policy Implications:

- The need for accounting, or other regulation, in an EU perspective
- The need for model development and testing
- The need for future research and research organisation

Sven Age Westphalen, from CEDEFOP in Thessaloniki, joined the workshop. He reported limited knowledge of human resource accounting beyond the community of enthusiasts. He discussed the implications for stakeholders, set out in a new report.

Gregory Wurzburg talked about work in OECD since 1992. Many companies waste money on training: they spend, but fail to manage the investment. This suggests that companies do not take training seriously. A conference was held in 1996, in the context of the knowledge economy, to see if the barrier to diffusion of new approaches could be overcome. It was concluded that financial reporting was not the best route. The Skandia approach was presented. The Norwegian Confederation of

Business and Industry favoured a collaborative approach, using a third body setting standards on disclosure of non-financial information. This would help disclosure, and help investors. Since 1996 the work has brought in committees concerning human resources, industry and capital markets. In mid 1999 there is to be a symposium, bringing in researchers from company-based work, looking at specific indicators, and assessing the feasibility of robust indicators that can be audited. Is it possible to do this on a standardised basis? The symposium should bring in government officials concerned with market regulation, looking at the needs from a political perspective. Other approaches to disclosure will be considered, including benchmarking and Investors in People, noting preferences of some companies for confidentiality. Lessons will be sought from environmental accounting and financial reporting. OECD needs a political mandate in order to proceed: the US SEC objects to others active in its area of concern; the US Federal Accounting Standards Board is starting work. The outcome should be a clearer view on non-financial reporting. The conference programme is being refined.

Sven Age Westphalen reported on work at CEDEFOP on mission effectivenesss, since 1992. The financial aspects of training in member states are being prepared. Work is needed on measuring techniques and frameworks, in association with EUROSTAT and DG-XXII. This is to be tested for the next three months, with international researchers. The target is measuring effectiveness of training, in terms of return on investment. More and more human knowledge is informal, and hard to measure. Human resource accounting is emerging, but how effective is it, and what are the political implications? What is the vocabulary used? Can we advocate a more consistent terminology? What does the discussion cover? Can we harmonise? The next step is a terminology book.

Stefano Zambon asked about the framework for measuring effectiveness on a true financial basis, and asked about expert involvement, which include **Ulf Johanson** and Ann West. Why are the European Accounting Association not involved?

Alfred Gutschelhofer asked about accounting standards setting committees, and saw signs of resistance to change. **Gregory Wurzburg** confirmed that there are problems, and noted that IASC are taking several years. There can be good reasons for taking time. Whether this leads to a change of practice is not clear, but it will take at least 15 years. Non-financial information can be important if it is in a standardised format. The accounting bodies have become more receptive. **Stefano Zambon** suggested a possible symposium at the European Accounting Association conference, of which he is chairman.

Sven Age Westphalen then spoke about work led from DG-XXII, who have been looking at the financial aspects of Computer Based Training. EUROSTAT is about to work on statistical indicators for intangible assets.

Gregory Wurzburg recalled an OECD report on guidelines for collecting statistics, and subsequent work on intangibles and indicators. Even with good surveys, companies are not in a position to provide the information, which they have had no reason to collect. The work we are discussing is upstream of current work at EUROSTAT and elsewhere. **Ulf Johanson** reported on the availability of such data from Dutch firms. The effort could be in vain, as company focus changes. **Sven Age Westphalen** noted that Sweden is a good starting point, but member states such as the UK lack the core information.

Laurie Bassi is addressing the same set of issues at the American Society for Training and Development. How can companies be persuaded to participate in a good survey? The largest project at ASTD concerns education and training, working from 1991 with large multinationals who realised they were spending large sums but knew little in detail. The Benchmarking Forum has developed common definitions and metrics for spending on education and training, enabling benchmarking and meeting internal management needs. In 1996 **Laurie Bassi** joined, and made the methodology more widely available. At the same time she worked with a subset of the group to develop benchmarkable outcomes in education. These were then made widely available. The work has been laboriously detailed, but has been valued by participating companies. In 1998 the programme has included 400 companies outside the USA, including 58 from Europe, and initial reports are now being sent, giving firms feedback to motivate their continued involvement. The more experimental area has a smaller scale of involvement, and will take some years to unfold: it involves new evaluation strategies. These projects provide the core data, with market information on publicly traded firms. The research objective is to look at linkages over time between investments, intermediate outcomes and financial performance. This is difficult.

Sven Age Westphalen talked about the treatment of intangibles as investment, rather than cost, by DG-XXII. Their commitment is to increase investment in human resources. Measuring techniques that expose excessive inefficiency could lead to restructuring or reduction of such investments. They are considering tax regimes in the member states. **Alexander Kohler** is preparing a report for DG-XXII, and there are ongoing debates on harmonisation of tax regimes. There is also work on training accounts.

Stefano Zambon noted the importance of fiscal implications of changes in accounting procedures. Capitalising costs may raise tax liability, which would have particular impact in countries where tax is linked to profit and loss. There are answers, such as making items deductible. **Gregory Wurzburg** asked what the objective is, and noted that some changes in accounting, linked to tax, could lead to reduced training investment.

Ulf Johanson described the MERITUM project, involving six nations, looking at the effects of misreporting intangibles in firms. Other work looks at the importance of management and control, intangibles, and of auditing. Policy implications will be considered.

The Need for Accounting, or Other Regulation, in an EU Perspective

Competitiveness is a central issue for the EU, together with the quality of working life. **Guy Ahonen** reported on Finnish working groups, which conclude there should be no mandatory regulations, but models for good practice. Information is needed, concerning the problem of the ageing labour force. There are problems with early retirement, presenting economic challenges. Better disclosure would be one way forward, improving working life. The timing of is normally due to market failure, as with the Nike case.

Laurie Bassi argued that there is a case for declaring market failure in the development of portable educational skills. With less than perfect information and aversion to risk, markets do not work. Without full information, and with risk-averse behaviour in firms, under-investment is likely. Intervention would provide a public good.

Gregory Wurzburg noted the policy commitment to lifelong learning, and the need to improve access to and returns on investment. **Jan-Erik Gröjer** noted a potential role for regulation. **Gregory Wurzburg** looked at paying for lifelong learning, seeking to lower costs of capital to firms. **Per Bukh** discussed the motivation behind the work in Denmark, where some officials are in favour of regulation, whereas companies oppose the idea. He did not anticipate a specific framework as an outcome. The question is how the information is used in companies.

Alfred Gutschelhofer saw political parties as raising problems, with cost-cutting becoming popular. Something has to be done. Retirement and lifelong learning are extremely expensive. People work for as little as 20 years between university and retirement. The life experiences of males and females are different. Companies need to take a different role when recruiting staff from university and passing them on to retirement. It is important to be able to account for human resources. There needs to be dialogue between public and private sectors.

Robin Roslender discussed employment reporting as an extension of corporate reporting, with the workforce as the audience. The UK government are still considering requirements for corporate reports.

Alfred Gutschelhofer indicated that companies know something must be done. The labour market is not working well, and it is hard to recruit the necessary people. We have high working costs. **Thomas Gunther** talked about unemployment in East Germany, and training of people who will never re-enter the labour market, while skill shortages remain. Tax problems would be serious for Germany.

Stefano Zambon noted a lack of awareness of the issues in Italy, though unemployment is a major concern. Recent measures have included tax allowances for recruiting new staff. If we agree that Human Resource Accounting is important, it must be made politically relevant and interesting, linked to tax incentives for companies. The issue is political, not just accounting. Training is a problem: how do we invest, how do we measure the success? Italian trades unions have yet to be interested. It will be hard to bring such information within the current system.

Robin Roslender described the UK situation, where financial accountants rule. Short-termism must be overcome. The UK is a "now" culture, and the new government has shown no real concern for the quality of working life. Trade unions, he argued, have not been interested in these issues.

Laurie Bassi argued for the importance of the market. Alignment of objectives can develop in the field of training. Regulation only arises in cases of market failure. How could information be used to help labour and financial markets without involving accountants? Where skills are scarce, companies may change their approaches.

Gregory Wurzburg saw three steps of argument.

1. What is broken? Market failure considerations must go beyond under-investment in training and intangibles. Alternatives to training may have more measurable outcomes. If something is broken, such as in capital markets, there can be too much capital intensity.

2.   What remedies can there be? Better information is important: how can it be economically feasible? Quality of working life, improved markets and internal management. Disclosure of information about training needs to be seen as improving the outcomes.

3.   Then the question is of financial or non-financial, mandatory or voluntary. Some new rules should be applied to non-financial information. It needs to link to financial information, and connect to established accounting procedures. In the US, if it is not mandatory, companies do not like to report it, as it may be used against them by shareholders in a law suit. Short-termism is partly about managing what you can measure: the current debate is about dealing with a wider range of information.

Sven Age Westphalen noted that training is not just about productivity, and will not all give financial returns. There are short-term and longer-term issues. Human Resource Accounting in a non-financial statement could be helpful. Trade unions are changing in role. He discussed the relevance of the ISO standards approach, voluntary and market-based. An alternative is the voluntary rewarding approach of Investing in People. Green accounting offers a model of a compulsory minimum approach. Body Shop uses this as part of their profile. The alternatives need to be discussed with the social partners.

Stefano Zambon asked if Swedish unions would take a lead. **Gregory Wurzburg** asked if they had operated internationally, and noted that unions in most countries have been slow, with Denmark as a possible exception. **Laurie Bassi** reported that this is a critical issue for trade unions in the USA, whose survival is threatened.

Per Bukh reported on Danish trade unions, who look at Intellectual Capital Accounts with some scepticism. He highlighted the links between training and competitiveness, as motivating companies across the sectors in Denmark. Involving trade unions and the workforce is good for the company. Intellectual Capital Accounts are an instrument for creating a culture in the company. There could be problems with legal requirements, but suggestions are important. OECD recommendations would carry weight. **Jittie Brandsma** endorsed this view from the Netherlands. **Gregory Wurzburg** recalled the Norwegian voluntary standard approach, which had been effective. **Per Bukh** referred to social reporting in Denmark.

Guy Ahonen commented on trade union involvement. It is not hard to involve leaders, but it is not clear that this is helpful. First it is being branded as good business. He does not include it in bargaining, but as a financial and legal market structural issue.

Robin Roslender talked of what is "good for business" in the UK. A new approach would need to be sold as relevant to management accounting. Work is beginning, but it is focusing on the amount of money spent on human resource management. The challenge is for it to be strategic. Financial markets are not indicative of good business, but of success from the point of view of investors, which is not the perspective of human resource accounting.

Laurie Bassi agreed that involving the power of the unions might not help the cause. However, unions control large pension funds, which they would like to invest in companies which support employment growth and enlightened practices. This could be a route to pursue in Europe, where applicable. **Stefano Zambon** said this was less applicable in Italy and Southern Europe. **Thomas Gunther** said this is now an issue in the former East Germany. **Guy Ahonen** said there were only two or three such funds in Finland, and one is setting up a risk management scheme, for which they have asked for a human resource report (in return, companies have reduced rates). In Denmark **Per Bukh** noted there has been a debate about pension funds and ethical issues, rather than Human Capital Accounting. The funds are owned by the workers, but with external professional management. **Thomas Gunther** discussed risk: German companies must set out details of their risk management systems. Every executive is held personally responsible for having a risk management system.

Jittie Brandsma noted that the Dutch government is reluctant to impose regulations. The need is to stimulate effective investment, but employers will oppose regulation. Do we have enough knowledge to enable us to impose regulations? Large companies can deal with these questions, but it is difficult to involve SMEs. The additional complications could discourage them from training.

Sven Age Westphalen described problems of SMEs and family businesses in Greece. **Gregory Wurzburg** noted the need for a language in which SMEs could discuss these issues with banks. **Jittie Brandsma** said that the issue concerns the imposition of regulations. **Gregory Wurzburg** and **Jan-Erik Gröjer** discussed different forms of reporting. **Ulf Johanson** noted the importance not just of capital markets, but also of working life. Reporting could be of indicators.

Stefano Zambon noted the need to talk to banks, fundamental to these issues of financial reporting in most European countries. He distinguished global players from SMEs. Global players are moving to international accounting standards. There must be incentives for the SMEs when they invest in people, as there are for employment and technology. However, trillions of lira are devoted to training, but either the money is not used or it is used without vision.

Jan-Erik Gröjer asked what the process would be if a recommendation was made. Should it be an audit process, or is this a bad approach? **Thomas Gunther** takes a narrower stricter view of auditing. Each country has a different system of companies, including large unquoted companies. **Stefano Zambon** endorsed the view: "the market" is not simply a matter of capital markets. In Italy 200 companies are quoted. **Thomas Gunther** echoed this. **Laurie Bassi** asked about the percentage of employees in quoted companies: small in these two cases. **Gregory Wurzburg** saw auditing as essential. **Thomas Gunther** cited auditing as with environmental and ethical auditing, or certification. **Laurie Bassi** argued for the importance of auditing of information concerning the future. First, we have to establish what is important. **Guy Ahonen** noted that everything is possible in Finland. KPMG have done their first human resource report audit.

Robin Roslender noted that much of company reporting is not audited: advertising is one purpose of human resource accounting or employment reporting. **Laurie Bassi** recalled that there are laws against false advertising. **Robin Roslender** declared that most cases against auditors do not go to court. He is reluctant to rely on auditors.

Stefano Zambon raised the comparative perspective. Auditors talk of a true and fair picture, and there is a problem of credibility within and across countries. The scope of accounting is being extended, covering assurance. We should indeed be cynical, but we are overlooking company structures, and a possible new Directive, moving towards a more German structure, with a supervisory board, including representatives of shareholders and workers.

The Need for Model Development and Testing and The Need for Future Research and Research Organisation

Per Bukh summarised conclusions from Denmark. We need a solution that is easy to use, but there needs to be debate. Models need to be tested, as the Danish project is doing. There is not yet consensus about the concepts on which to build. Human Resource Costing and Accounting is right, but something simpler is needed at present, such as an indicator approach. Companies that cannot use indicators will find financial figures even harder. Verification can be through certificates. More research is needed on other approaches. **Ulf Johanson** asked whether construction of the model is research or practical. How do companies deal with these issues? Do we want standardisation, or diversity? Is it too soon to standardise?

Laurie Bassi agreed that we do not yet know what is most important, though US research points to activities producing excellent performance. The choice of indicators should be a collaborative process guided by researchers. **Gregory Wurzburg** saw indicators as signals to the world outside the company. He looked at ways of doing this, such as through the ASTD benchmarking forum. Another approach is to use confidential data to check against, for example, Investors in People. **Ulf Johanson** envisaged large databases, managed by the third party doing certification.

Sven Age Westphalen liked the Danish involvement of the public. At some stage there would be focus on a more limited group, with a more structured approach. It should not just be a management tool, and public involvement, including the trade unions, is important. **Laurie Bassi** looks for lifelong education tools, and needs to involve the employer. Because of outsourcing of education and training, it appears that ASTD have a tool to evaluate the quality of vendors.

Guy Ahonen described human resource disclosure in Finland. Concepts and indicators will vary between companies, addressing similar elements of intangibles. We need to consider human resource profit and loss accounts. There needs to be research on competence inventories, arriving at representation in figures. Benchmarking saves revealing delicate information.

Thomas Gunther wants Human Resource Value Accounting, but the problem is measurement, and the choice of indicators. He discussed strategic management concerns and strategic key factors. What should be the relationship between human resources and intangibles? There needs to be a theoretical framework, dealing with innovation, research and development.

Jittie Brandsma was interested in exploring causality, linking investment in training with output indicators or productivity increases. Organisational structure may be of particular significance.

Jan-Erik Gröjer asked about the scale of the population needed for viable research. **Laurie Bassi** declared that panel data is needed. **Per Bukh** asked about the use of qualitative data, based on study of companies over time.

Stefano Zambon distinguished the institutional level for "ignition" purposes, bringing together the key actors, from the research level. The construction of indicators should not be left to academics. Testing should involve political agreement at a supranational level. There is a danger of superficiality, as with industrial performance dynamics, some years ago. Countries vary. He argued against simplistic visions. He cited *In Search of Excellence*, where all the examples of success have since gone bankrupt.

Gregory Wurzburg wanted to avoid just another social science research project. Users of the new kinds of information will look at the academic research, and at what other investors and companies are doing. The field is vast, and it is hard to identify a particular approach to follow. After the 1996 conference OECD discussed sending out research groups with a common design, but there are already groups working in their own ways, which we should track. The project needs political, business and research credentials. It is important to monitor the impact of the release of particular kinds of information, such as on customer satisfaction, or the Baldridge Award. Similarly, we should monitor the impact of downsizing on share value.

Ulf Johanson asked about the banking system. **Gregory Wurzburg** would like post mortems on bankruptcies. **Jan-Erik Gröjer** said that some of this would be addressed in the MERITUM project. **Guy Ahonen** emphasised the importance of banks for SMEs. **Thomas Gunther** was asked about the importance of banks. Bankruptcies were linked to shortages in management skills, among owners and employees. Human resources appear to be the most important item.

Robin Roslender argued for a clearing house for this kind of work, which perhaps should be in Stockholm, at the Personnel Economics Institute which hosts the international journal. He wondered whether some other European institution was needed for Intangible Assets Research. There is relevant work in the USA, but no visible focus in Europe. **Stefano Zambon** reported interest at the European level, but there would need to be funds and support.

Jan-Erik Gröjer prefers a virtual organisation to empire building. When there is a product, then there may be a need for a new organisation. The workshop group could constitute a virtual institute. **Ulf Johanson** described conversations with universities and ministries. The problem is not money, but the nature of the virtual institute.

Gregory Wurzburg argued that standards setting is not the task of a clearing house, but of existing institutions such as the European Commission. OECD cannot support major ventures. Ideas need to be anchored to existing organisations. **Jan-Erik Gröjer** wanted provision for those who have no funds. **Stefano Zambon** repeated the distinction between research and practical activities, involving political support. He talked of a virtual institute to deal with intangible assets and would be exploring the potential for symposia at the European Accounting Association conferences in 1999 (Bordeaux) and 2000 (Munich). **Sven Age Westphalen** declared that CEDEFOP are happy to be involved with the virtual institute. **Robin Roslender** emphasised the importance of not for profit organisations.

Ulf Johanson noted that the importance of the workshop had been to explore the state of the art, and to find ways of moving forward. He and **Jan-Erik Gröjer** would be producing a report, and requesting comments. The resulting document can then be used in a number of ways.

Workshop Participants

Guy Ahonen, Swedish School of Economics and Business Administration, Finland
Laurie Bassi, American Society for Training and Development, USA
Jittie Brandsma, University of Twente, Netherlands
Per Bukh, University of Aarhus, Denmark
Leandro Canibano, Universidad Autonoma, Spain
Richard Ennals, Kingston University, UK
Eric G. Flamholz, University of California at Los Angeles, USA
Jan-Erik Gröjer, Stockholm University, Sweden
Thomas Gunther, Technische Universitat Dresden, Germany
Alfred Gutschelhofer, University of Graz, Austria
Ulf Johanson, Stockholm University, Sweden
Alexander Kohler, European Commission DG-XXII
Rainer Marr, Universitat der Bundeswehr, Germany
Maria Martensson, Stockholm University, Sweden
Jan Mouritsen, Copenhagen Business School, Denmark
Robin Roslender, University of Stirling, UK
Herve Stolowy, HEC School of Management, France
Karl-Erik Sveiby, Australia
Sven Age Westphalen, CEDEFOP
Gregory Wurzburg, OECD
Stefano Zambon, University of Padua, Italy

Reflections on the Workshop

The dialogue was stimulating, and concerned not just with academic theory but also with future practice. It was clear that current international accounting practices leave a great deal to be desired, but that there is no one single alternative model ready for introduction. National practices, cultures and economies have varied, but it was recognised that the global economy means that new international discussions are required, possibly leading to new conventions, which could also play an important role in future efforts for European harmonisation.

New perspectives emerged, with valuable insights into measuring the effectiveness of investments in training. It is established wisdom that training is "a good thing", but there has been surprisingly little work to date on measuring its effectiveness, which can leave training budgets vulnerable at times of difficulty. Training needs to be linked both to organisational development and to reformed accounting practices.

The participants were authoritative experts, and agreed on the importance of ongoing work. The second workshop from the same leaders, on Intangibles, may show the benefits of the increased international collaboration.

4. Policies for Occupational Health and Safety Management Systems and Workplace Change

The workshop was led by Kaj Frick of the Swedish National Institute for Working Life, and held on 21–24 September 1998 at the National Trade Union Museum, Amsterdam.

Abstract

The workplace, the workforce and the labour market of industrialised societies are undergoing rapid and profound change. These include physical changes, due to new technologies, changing patterns in labour force participation (most notably by women), changing demographics (notably an ageing population), changes to legislation, changes to organisational structures and changing forms of employment relations. These changes have significant implications for Occupational Health and Safety (OHS) management. Internal quality control of the work environment is being promoted as the new strategy to solve OHS problems. The workshop aimed to provide perspectives on this strategy, its interaction with other developments in an around the workplace, and its effects on OHS. There was a focus on the relationship between regulation, privately based promotion and OHS management.

The workshop was concerned with workers as the most direct beneficiaries of success or victims of failure with regard to OHS management, and with their right to a major say in such processes. Worker participation is integral to OHS management, or it amounts to little more than cost or risk minimisation. The workshop was concerned with participation at workplace, industry, society and government levels.

Papers had been prepared and circulated in advance by the organising committee comprising **Kaj Frick, Per Langaa Jensen, Pascal Paoli, Michael Quinlan** and **Ton Wilthagen**. The workshop brought together many of the themes of Work Life 2000, and linked discussions from earlier workshops. At the end of the workshop **Anders Schaerstrom** gave an introduction to the SALTSA Joint Programme for Working Life Research in a European Perspective.

I. Roots and Origins of OHSM

Kaj Frick and John Wren
Reviewing OHSM at the Close of the 20th Century: Multiple Roots, Diverse Perspectives and Ambiguous Outcomes

Kaj Frick offered an overview of the field and the conference, looking at management systems for health and safety. He described a third phase of strategies, following the industrial phase and voluntary safety management. Reforms in the 1970s specified what was to be done, by clearer and stricter regulation. An enlightened approach has underpinned research and development in the industrialised world. What is to be done, and who should do it? There have been different approaches to worker participation. The 1980s brought the third wave, the politics of health and safety management. There was increased awareness of the costs of industrial illnesses. Another context was the debate on deregulation, in which health and safety management systems were seen as a panacea. Ideas were drawn from ecological and environmental politics.

There are now four general trends:

1. Marketing of voluntary health and safety management

2. Attempts to standardise and develop ISO guidelines

3. Government public regulation, such as the EC Framework Guideline 1989

4. Hybrids of voluntary and mandatory systems

It is not as easy as some had hoped!

John Wren set out horizontal and vertical problems. The OHSM paradigm needs to be challenged: is it a panacea? He looked at the arguments from **Tom Dwyer, Michael Quinlan** and others. Different methodologies are needed, and OHSM does not necessarily fit the needs of SMEs. There can be threats to resources of companies and countries. Self-regulation has been interpreted in different ways since the 1972 Robens Report. Is deregulation to be seen in terms of form, or of substance? How do we harmonise across countries and cultures? How does OHSM fit in with industrial relations? What are the resource implications? What are the priorities for resource allocations? What justifications are put forward? What accountability is there? How can OHS be integrated with environmental concerns? Do we need a fundamental rethink? Or do we need to develop more detail?

Tom Dwyer
Decline and Renewal: a Developing Country Perspective on the Industrial Health and Safety Paradigm

Tom Dwyer was concerned with action, participation, explanation, listening to those who suffer, and indices of health and safety. He noted the interdisciplinary nature of the conference, and perplexity about a world in complex change. He described the collapse of Cartesianism, central to the great Western European intellectual tradition, and the development of action as well as research.

Accidents in health are to be seen as outcomes of social systems. Development, and Third World development, have to be seen in a similar way. Brazil's environment is of savage capitalism, similar to the USA a century ago. There are three distinct systems: pre-industrial, industrial and post-industrial. Workers are heard less than in the developed world, and major accidents have followed. Pre-industrial workplaces escape state attention, and have to develop their own vision, which is of anthropological interest. Industrial management combines paternalism and authoritarianism. Questions arise about relations between government and occupational health and safety services. He cited participative approaches, and their regulation, before considering post-industrial workplaces, where safety and health measures are a precondition for production.

Today the actor and the system are dissociated, and there is no move to make major change. Professionals work without clear evaluation. He cited the work of **Pascal**

Paoli on change in the developed world, moving, with deregulation, back to savage capitalism. All of the current systems are subject to power relations. Participation is a concept in need of a theory, or it is simply ideology. We have to question the industrial paradigm of safety and health, as the great Western paradigm has crumbled. We have to redefine cause in terms of social relations.

Carolyn Needleman
Conflicting Frameworks for Regulating Occupational Health and Safety in the United States: "Cooperative Compliance" and the Struggle to Control Ergonomic Hazards

There is increasing interest in voluntary compliance. The Brazilian situation is based on great power imbalances. In the USA there has been experimentation with voluntary compliance, but there are problems. Events have moved fast in the USA, with new policies delayed in the courts, and continuing battles in Congress. Ergonomic standards have faced even more delays.

There are two styles of regulatory enforcement: adversarial and collaborative. One is based on amoral financial judgements, the other on persuasion and goodwill. Some combination of the two is required. In the USA, pressure on employers is required, and the labour movement is weak. The issues are targeting of compliance programmes, the implications for enforcement resources, and the consequences of combining approaches.

The Occupational Safety and Health Agency (OSHA) Consulting Service is available in association with the states, but faces legal problems of reporting (based on an adversarial rather than collaborative approach). The service benefits small businesses. The take-up of the service is low. The Voluntary Protection Programme is better known, where good practice is shared as a means of avoiding inspections. This targets the best actors. In California they have implemented cooperative compliance, again based on good practice. The "Maine 200" programme targeted sources of complaints, as does the national OSHA plan. There has been opposition from employers, worried about ergonomics control by the back door, resulting in the current court cases. OSHA is working on standards for health and safety for all workplaces: there is likely to be debate. The defence of ignorance of the hazards would be removed. The workforce are uncertain of collaboration, and critical of retreat from adversarial approaches. The government finds the OSHA process labour and resource intensive, and resources are lacking. OSHA has only $330m, compared with $7.8b for Environmental Protection.

Pascal Paoli
Working Conditions and Public Authority Involvement in a World in Flux

In the European Union, there are 18m unemployed in a labour force of 147m. The self-employed remain at 18%. Industry is 32%, Service 61%. There is increasing work with computers and with clients. Fixed term and temporary employment has increased, especially in the South. Entry to the labour market tends to be via temporary jobs.

Employers demand that precarious employment should not be discussed as a policy issue. Temporary and fixed term workers have worse working conditions, less training, and less consultation. The worst jobs are given to those with precarious status. Things are getting worse for many workers, with 10% suffering combined pressures. Work is intensified and unevenly distributed. Backache, stress, fatigue, muscular pain and headaches remain a problem. Women are an increasing proportion of the workforce, especially in precarious and part-time work.

Pascal Paoli outlined the limits of the traditional regulatory system of regulation, control and sanctions. Inspection levels vary, and internal control raises problems. Trade union membership is low in countries such as France. He discussed the challenges of multidisciplinarity. There are challenges to single discipline systems posed by e.g. stress. How do we evaluate risk? This should involve workforce participation, which is a political issue. The role of law varies, raising questions about public authorities.

How can we move from insurance to prevention? The 1989 Framework Directive requires annual evaluation of risk by employers. There are then technical and broader philosophies on the scope of such evaluations. There is a role for economic incentives, and for integrated corporate approaches. He considered new indicators and monitoring systems for the changing situation. The demands become ever more complex, but funding is being reduced.

Theme 1 Comments: Theo Nichols

Theo Nichols attributed greater resort to OHSM to fear of Japanese competition; the weakening of labour; a long wave of depression, with deregulation, privatisation and cost reduction; and commodification, even of OHSM, with a tendency to regulation.

We cannot distinguish consequences and origins, but the introduction of Human Resource Management, Total Quality Management and Just In Time raises the issue of Balkanising knowledge. There is a literature on empowerment and lean production. We are aware of changes in the structure of employment, with a move to precarious work and self-employment. Workers are vulnerable, to varying extents. There could be no universal prescription. The UK Health and Safety Executive have launched a new strategy: some firms are so small that talk of management, and of management systems, is useless. There are power relations involving purchasers of goods: our unit of analysis is the entire work system. Our understanding of health has to include freedom and dignity, which gives rise to criticism of some corporate systems.

II. How to Apply the OHSM Strategy in a Complex Labour Market

Michael Quinlan and Claire Mayhew
The Implications of Changing Labour Market Structures for Occupational Health and Safety Management

It is vital to consider changes in the labour market. The focus here is on health management systems. The main changes, in the view of OECD, are:

- More women in the workforce

- More youth
- Increased shift work
- An ageing workforce
- Decline in the proportion of males working full-time
- Increased outsourcing and downsizing
- Increased self-employment
- More small business
- Decline in job tenure
- More part-time and precarious work

We should expect these changes to accelerate. We need more data, with research reported on the OHS consequences of outsourcing. There are economic and reward factors which diminish OSH, including cutting costs, offloading risk, and long hours. Disorganisation is a consequence of ambiguous rules and practices, complex communication, and the inability of outsourced workers to organise themselves. It is increasingly likely that regulatory systems will fail. Minimum labour standards are not met, with long hours, low wages and child labour. Governments are affected by the fall of employment in large workplaces, with poor monitoring, and workers' compensation data provides inaccurate patterns as casual workers underclaim. Databases are often flawed because of their definitions of work. OHS services are affected. Governments are beginning to realise that there is an issue.

These changes are due to changing employment practices of big business in a neo-liberal environment. Employers do not want to address issues with potential large costs: they prefer to offload expensive activities. Leaner production is not a route to improved health and safety. Unions are weakened by these changes: they need to be involved in monitoring and in change.

Claire Mayhew summarised recent research, producing a mass of data. Injuries and fatalities have been shown to be double or treble those of standard employees. Contingent workers have a low level of legislative knowledge, a resistance to the regulatory framework, and a reluctance to claim. Government intrusion is resented. There are signs of hope, such as in recent work she reported from a multinational food company, which demonstrates understanding of the law and risk assessment, applied to sub-contractors as conditions of contracts. This is not typical of the fast food industry. She also cited requirements of small builders to undertake risk assessments, resulting in improved OHS outcomes. The evidence is coming together, including new risks in contingent work such as for casual workers in the fast food industry, who may have to deal with drug addicts. Women and young people take the majority of these jobs.

Felicity Lamm, Joan Eakin and Hans Jorgen Limborg
International Perspectives on the Management of Health and Safety in Small Workplaces

Felicity Lamm discussed OHS and small businesses. Small businesses lack management and training skills, and awareness of occupational health and safety. Employment relations are peculiar: employees are typically not unionised, female, and with English as a second language. Complaints are hard to make, as the complainant can be identified. Illegal migrants are employed and exploited. Labour-only subcontracting causes problems for the individual. Relations with large businesses and regulatory agencies cause problems. Small businesses are not miniaturised large businesses, but have their own problems, requiring different approaches.

Hans Jorgen Limborg addressed the Danish Dialogue/Consultation programme in the Danish Occupational Health Service. The dialogue network has run since 1993, with an impact on occupational health. The 1997 survey showed that small businesses are on the agenda, but with no clear methodology. The report became a handbook, based on "10 Commandments".

1. Personal contact is essential

2. Do not address working environment problems

3. Accept that survival is the key problem

4. Be ready to be involved in action

5. Highlight what has been done

6. The solution is integrated in the problem

7. The owner is the key person

8. Be available

9. Assign responsibility for maintenance

10. Build on local practice

Joan Eakin outlined the Canadian Community Development approach, based on the Canadian Safe Communities Foundation, with private corporate funding, small municipalities, cross sectoral public safety focus and a workplace component. It combines

- Mixed public–private resources

- Community-based, sociological approach

- Accommodative approach

- Economic motivational approach

- Cultural model

The project has been successful, spreading to 11 communities, and producing reductions in injury costs. She asked whether this approach to linking OHS to community safety deflects attention from OHS? Does it enlarge corporate control? Does the soft touch approach depoliticise the issue? What are the implications of a community development approach in large urban settings? What are the implications of piggy-backing? Can OHS staff do effective community development? Does the approach encourage non-reporting? What are the implications of private sponsorship? Does it deter labour? What are the implications for new OHS work, such as contingent work? To what extent is it generalisable?

Tore Larsson
Consequences for the Regulation of Ohs Management of Changing Workplace Structures

The normal focus for **Tore Larsson's** research is development work in corporate organisations. He presented an analysis of the changing workforce structure in Sweden. Out of 195,000 firms, 167,000 are non-corporate with fewer than 50 employees. There are 1,600 public employers, 3,600 corporate employers of more than 50 employees, and 24,000 corporate employers with fewer than 50. Thirty-seven per cent are public employees, with new structures of employment being introduced. Thirty-three per cent are in corporations with over 50 employees, and only 18% are in non-corporate small firms.

Traditionally the employer was seen as God, but in late modern society people will not accept absolute rulers. OHS institutions were introduced a century ago, and the principles remain valid. The responsible employer idea was pre-modern in origin. Society's control through legal means continues. As a modern example he considered Akzo Nobel, and the developing of coating activities over 30 years, with the rationalisation of the responsible employer. Recent developments have been a combination of deregulation and institutionalisation, with the reduction of risk. An analysis of industrial accidents in the 19th century showed deaths and damage, giving scope for widows to seek compensation. There had been no functioning credit system, but it was clear that companies could not risk more than their assets. Workers could move their energies to developing systems of support, not resentment. The labour inspectorate was also an instrument of control. The abstract employer was accepted as responsible. With divisionalisation, employer responsibility could be decentralised. The next stage was legal persons as sub-units, with corporate executives outside legal responsibility. This raises problems for occupational health and safety, with centralised decision making but distributed responsibility. Information systems had to be introduced which did not give full information on health risks. New labour law is needed, offering new ways of regulating the work environment.

Theme II Comments – Ton Wilthagen

Ton Wilthagen responded to the set of papers. OHSM systems depend on stability, transparency and maturity. The papers suggest that the preconditions are not being met. The labour market is being destabilised. Large firms are falling apart, posing a challenge to management systems. The diffusion of management responsibility adds to the problem. Small firms have never taken on the systems approach. Prospects in developing countries are bleak.

However, it is possible to be optimistic: for example part-time work is being recognised. We need to complement management systems with other strategies.

- Changing the law on responsibility; financial incentives for employers. Social security policy can help create better working conditions.

- Supportive or first aid strategies should enable firms to become stabler and more transparent. Dialogue and community approaches can be used.

- More fundamental strategies try to extend the coverage of OHSM systems in a more radical way. This means reforming law and regulations, encompassing atypical workers. The employee is dependent on the employer. This could be replaced by economic criteria. There could be a price to pay for bringing people back into labour law: it contributes to a flexible economy and to the security of workers. New entities can be created, made up of a number of small companies, with rights and obligations. In turn, this requires a broader perspective: social security, company law, labour law are all involved.

III. OHSM as Part of a Voluntary Management Process

Peter Dorman
If Safety Pays, Why Don't Employers Invest in It?

Peter Dorman identified the managerial perspective, seeing OHS as a management problem; and the market perspective, in terms of benefits and costs, where OHS is a problem of cost externalisation. He located the debate in a neo-liberal context, affecting OHS together with other policy areas. The push comes from ill-informed critics, who do not realise that market approaches did not work in the past, thus the move to regulation. It is worth looking at ways of improving the market approach, within a wider OHS system.

The OHS cost accounting problem was considered. He identified two problems:

1. Inside the firm, costs are transmitted without taking OHS into account. It is hard to identify the hidden costs. This is part of the challenge of human resource accounting.

2. Incentive design needs attention, and can be distorted. Reporting systems have been affected by the push for incentives.

Another issue concerns control. **Peter Dorman** served on a federal panel on child labour, and addressed issues of secondary (casualised) labour markets. This is becoming more prevalent, and reflects employer choice, in the direction of cost minimisation. Incentives are weak if employers have preferred secondary labour. A systemic shift is needed. Incentives can be counter-productive. It remains hard to account for the costs of OHS, which would be required for an effective incentive approach. We have to distinguish injuries and diseases. We cannot solve the human cost problem.

He called for accounting reform, but noted that the problem will continue as long as workers are seen as external to the firm. He called for better incentive design, and policy integration. He acknowledged the limits of the market approach, in the area of disease. Attention must be given to non-economic cost and non-economic values.

Kees Le Blansch and Annette Kamp
Participation, Prevention and Control: Linking Management of OHS to Environmental Protection and Quality Management

The paper reported work from the Netherlands and Denmark. Danish work was presented first, describing the integration of OHS and the environment. Is this a strategy we should support or resist? Linking OHS and the environment is given high priority, with a move from command and control to regulation. There are material links between the fields, and other stakeholders must be considered, including customers and community groups. What have been the arguments for linking the fields? It could raise the profile of OHS, revitalising activities, strengthening employee involvement in environmental work, and leading to more effective prevention.

Combining the two systems involves promoting standardised management systems, following the model of quality management. We inherit from quality management an uneven mixture of Taylorism and continuous improvement, with an emphasis on management control. It uses the technical–rational paradigm, with assumptions regarding the nature of knowledge. From quality management and the environment come transparency and options for new actor groups. Modifications are required, changing priorities, views of knowledge, and roles of actors.

Danish work reported has been in high trust companies, looking at organisational change. Top management looked at the environment, while OHS was for middle managers. Environment was seen as technical, while OHS involved workers' consent. In OHS the trends depended on industrial relations trends, either towards professionalisation or direct participation.

Kees Le Blansch described his own Dutch research, on employee participation in environmental issues, involving OHS. Different kinds of companies had to be distinguished, where the environment was a market factor, where management was formal, etc. Another project compared industry sectors: traditional, primary process and high tech. A further project looked at EMAS and ISO certification. He considered the actors involved, the nature of the management process, the strategic interest of issues, technical complexity of debate, and the level and nature of participation. The links become weaker the stronger the emphasis on the environment, and the integrated approach is often little more than window-dressing.

The research partners met for the first time at this conference, and their synthesis sees management systems as a means of getting managers in control of environment and OHS issues. This depends on the management's motivation for each of these causes, and the views of market stakeholders, with regulation standards. The fit is easier in environment than in OHS issues.

Klaus Nielsen
Organisational Theories in the OHMS Systems Approach

Klaus Nielsen sees large companies as an enlarged version of small companies. He gave an historical account of the US and Swedish background in classical management theory (different but both inspired by Taylorism), safety management, modern human relations and post-modernism, and the ISO system of management standards, where universalism rules unchallenged. He also considered experience of internal control in Norway, showing how tools can become goals in themselves. Workplace assessment was seen in legal, expert and subjective terms.

Preben Lindoe
Integrating Internal Control System Into Business Development

"Enterprise 2000" is running in Norway, and provides the context for the discussion. It was initiated by Bjorn Gustavsen, and is based on participation. Action research brings together international management concepts and the Scandinavian model of industrial relations, with the goal of leading to new forms of democracy and cooperation within enterprises and increased efficiency. This involves benchmarking the Scandinavian way, looking around Europe. The oldest network is TESA, and the newest is TfS.

The health and safety emphasis has increased since the Bravo accident, with onshore work in aluminium, food and healthcare. Internal control regulations were modified in 1997, suiting smaller enterprises. Implementation in Norway is under way. The arena is complicated by quality management systems working alongside OHS, in a context of deregulation. The enterprise is linked to the market, with a reduced role for legislation. ISO 9000 certification, Product Control Acts, ISO 1401, EMAS and Internal Control Systems of OHS all come together, working from different starting points.

North Sea operators determined a target cost reduction of 50%, changing methods and procedures. Participation is covered by Work Environment Acts and other legislation covering all levels in the enterprise. This is being studied in the Aker Stord yard. Representative bodies were linked to TQM organisation, via the departmental council. There are implications for the network of suppliers, and for the roles of action researchers, who also operate as consultants. How can learning among small companies be organised? New forms of conferences are being organised. It is instructive to compare networks.

Andrea Shaw

What works? – Strategies Which Help to Integrate OHS Management Within Business Development, and the Role of the Outsider

Andrea Shaw is a consultant in the Australian state of Victoria, dealing with OHS and workplace change in a number of sectors. She is a reflective practitioner, seeking to create real change. It is not just about zero injuries and accidents, which can be achieved by manipulating data. She wants a workplace where people are central. There are multiple perspectives. Her approach is based on a tool "ASSET". She spoke about the physical reality of the meat works, and the power of the controller of the chain carrying the carcasses. It also defines what counts as work, so the idea of leaving the chain for discussion does not work. The symbol has to be confronted. The technical and the social have to be redefined.

Andrea Shaw declared that every organisation has an OHS system. It is a matter of making it work better, and integrating it into decision making. We are creating a myth of a safety management system, replacing the careless worker. The formal ISO system makes no observable difference when you look at high performance. It is the symbolic and political dimensions that are important. The ISO based approach in Victoria has accompanied a rise in serious claims. The systems approach supports an individualised model of risk, and sophisticated remote control. The blame can be passed back to the worker if safety accreditation has been achieved. Regarding targets, the outcomes are contestable and ambiguous, and claims can be manipulated. She reported work on performance measurement, in particular in the mining industry.

She described a small country meat works, owned by husband and wife. The wife noted the personal impacts on their employees. There were issues of participation and risk management, and organisational issues in the context of change. The workforce joined in and changed the design, which worked, to the surprise of the designer.

On contract management, she saw some cause for hope. Public utilities have been obliged to contract out, and local residents have come together, seeing the case for doing work oneself rather than contracting out. This is possible because of the Health and Safety laws for contracting in Victoria. The issue is power and control, and not just money. She concluded with reflections on the role of the outsider, and worries about the shift from regulation to consulting. The role is to make the contingency of the current arrangements transparent, opening opportunities for dialogue. There are no ready answers, but scope for encouraging people to find their own answers.

Gerard Zwetsloot
Reviewing the Debate on International OHS MS Standards

As vice-chairman of the Dutch standardisation committee on OHS management for the last five years, **Gerard Zwetsloot** had practical experience to report. His interests are in knowledge and practice in improvement of the work environment. He prefers to talk of management systems, rather than programmes. His consultancy includes work with small business, and with DG-V of the European Commission.

Governments have promoted self-regulation, and this includes standardisation, with ISO 9000 often seen as a success. Standardisation is based on consensus among stakeholders, producing standards (specifications, meant for certification) and guidelines.

Main topics on international standardisation are:

Relations with regulation: how should it affect current legislation? How should it impact on the debate on deregulation and privatisation? What about enforcement strategies, and the impact on social security systems? A number of perspectives are involved.

The value of certification: added value diminishes over time; external verification has added value; certification is no objective on its own. When there is a clear market demand, certification cannot be avoided.

The involvement of workers: standardisation has never dealt with political processes. There are good and bad examples of worker involvement (the Dutch involved the unions from the start, the Greeks did not involve unions and Australian unions responded negatively). There are enormous differences in the world in industrial relationships.

Unions face new dilemmas.

- *The impact on international trade*: barrier to trade or opportunity. Opposing perspectives for international and local small companies, rich and developing countries.

- *Major incentives for companies*: via governments, advantages for certified companies? Market advantages? Insurers? Financial benefits? Image in the labour market? Is this limited to certified management systems?

- *The limits of standardisation*: OHS is not technical, not collaborative.

- *Limitations of EU policy*: little experience of OHS or MS, little liaison with environmental systems.

Gerard Zwetsloot then presented comments on the ISO 9000:2000 series draft. There has been a shift towards a more process oriented approach, away from Taylorism. He reported on the stakeholders at the ISO process. You have to be there in order to make a difference. There are risks of future conflicts. Standards do not produce uniformly successful results. ISO 9000 was intended for control, and not for improvement. Companies need external confirmation of their performance in order to gain credibility. Standards need to include an assessment of performance, checked by verifiers. Standards can be seen as helpful in terms of the labour market and image.

Theme III Comments – Felicity Lamm

Felicity Lamm identified particular themes:

- Voluntary self-management of OHS: the myth of the management system? Government and management driven, unions in reactive role, where participation is a management prerogative, reliant on good will.

- The integration of OHS into management systems (quality systems) and wider issues (environmental issues). Policy must link with practice.

Issues include competing standards and competing interests. There needs to be a compliant population and a supporting culture (international, national, industrial, organisational and in the workplace). There are differences between employee participation and quality management, involving conflicting paradigms.

She set out to construct a framework, contriving ground rules in an anarchic situation:

- Codify and standardise the process and link it to economic benefits

- Borrow management principles, and apply them to OHS

- Broaden the boundaries of OHS by making linkages

- Adopt participative approaches to managing OHS

The debate concerns power relationships, and the translation of economic and social principles to health and safety. Rhetoric does not always match practice.

IV. Implementation of Regulated OHSM for all Employers

Neil Gunningham and Richard Johnstone
The Legal Construction of OHS Management Systems

Neil Gunningham declared that enterprises can be seen either as leaders or laggards. We should be moving more enterprises to the role of leaders. A management systems approach can bring firms up to a given standard, and beyond, through continuous improvement. The challenge is to build on the strengths of both systems. Voluntarism has not worked very well over the past century, and we need to consider the mandatory approach considered in Norway. There can be a variety of problems and risks. The recommendation was more modest and incremental.

Lessons can be learned from EMAS and experience in Australia. A twin track approach is suggested, nudging those on the margin into a management systems approach. Bounded rationality and imagination need to be overcome. This can mean resources burdens for regulators, unless system-based firms are left to regulate themselves. The challenge is to free up resources without falling into deregulation. The example was of concrete life-vests: the process could be fine, but the outcome fatal.

Third party oversight owes much to EMAS. Third party audit and disclosure are required. Triggers are needed for government intervention. Tripartism is fundamental, with worker participation at every step, and enforcement by unions. Cheating will take place, and breach of trust must lead to legal action, via fines and incapacitation. This is not an argument for deregulation, but for redeployment of resources. There is a danger of abuse, but compliance is required from self-declared leaders. There may be too few participants, judging by precedents in the environmental domain. The alternatives are grossly imperfect, but they may be better than the *status quo*. Credibility will vary in different cultures.

Richard Johnstone discussed prosecution of regulated firms, and of voluntary firms that broke trust. He introduced the enforcement pyramid. Prosecution needs to work in this context, dealing with health and safety. The traditional approach can be criticised as being reactive to events, with the court dealing with events not the system. This makes it easier to shift the blame to workers or suppliers. The event is shifted to the past. Fines are generally too low, and signal that the employer can buy his way out. There is no influence on non-financial values, and often the fine can be passed on.

How can it be made more effective? It should be a deterrent, facilitating reform and rehabilitation. He introduced a set of principles: consistency, covering all duty holders, proper integration, integrating mainstream law, systems-based approaches, tough sanctions, union prosecutions, publicity of outcomes, and transparency of guidelines.

Karl Kuhn
Policy Strategies in Germany for Integrating OHS in Management systems

Germany has a voluntary approach. **Karl Kuhn** summarised German reservations to ISO standardisation, without guarantees of improvement. We lack evaluation evidence of the success of internal controls as used in Norway and Sweden. There are questions of balance between the state and labour inspectors. The situation of SMEs is particularly important. Germany has 35m employees in 2.5m companies, but only 1500 have more than 1000 employees, and 99% of companies are SMEs. Front-runner companies have OHS strength, but there are back-runners and an Eastern industrial desert. Industrial relations are based on consensus, with co-determination including OHS, high productivity and high salaries. There is great variety in management systems, including quality management and integration, benchmarked by big companies. ISO 9000 standards are widespread.

There is consensus on the use of OHSM, but the issue is how? There are controversies about demands and prerequisites, applications and effectiveness. We can compare international systems of different types:

- Audit systems (oil industry)

- OHSMs oriented to ISO 9000 (Spain)

- OHSMs oriented to Plan–Design–Check–Action cycle (GB, Ireland, Australia, Bavaria)

- OHSMs as element of meta-structure

Weak spots in these systems are that psychosocial factors are not included, SMEs are omitted, and there is an absence of worker participation. Systems are impractical, lacking guidance, transparency and motivation. It is important to collect more evidence. Users need evidence for their own situation: one standard cannot fulfil this need. The more concrete the system, the larger the scale of details. This in turn makes the system less attractive, reducing freedom for users.

He set out framework conditions. Systems should be voluntary, take into account SMEs, and not require external audits or certification. Costs must be acceptable, supervisory authority should not be jeopardised, and OHS should be included in quality management.

The next steps should be integrated MS as an option, with a platform to be developed by social partners. Discussions continue about regulation and deregulation. Employers want less administration, but those with risky products demand control of risks.

Per Langaa Jensen, Alex Karageorgiou, David Walters and Ton Wilthagen
Risk Assessment as a Means for the Management of OHS Experiences from Four EU Countries

Per Langaa Jensen introduced the European Union system of decision-making and implementation of Directives, many of which concern the social dimension. Two objectives have been the single market and the social dimension, covering working and living conditions. The key directive is 89/391 on Health and Safety at Work, setting the framework for the Member States. All employers are required to make risk assessments.

David Walters outlined the situation in Britain. The situation has been influenced by the extreme politics of the period. What do we mean by risk assessment? There has been an obsession with the methodology of inspection. Overall considerations of health and safety management have had a minor role. The Anglo-Saxon approach to occupational hygiene continues. The EC provisions were implemented as regulations in 1992, followed by worker participation in 1996. The Health and Safety at Work Act of 1974 remains the core, and EC provisions were inserted. However, the regulations were not taken as law, but as approved codes of practice. The Framework Directive talks about integrated prevention, which is absent in Britain. British regulations talk about being "reasonably practicable", thus allowing cost–benefit discussions, contrary to the intentions of European law. There has been no broad evaluation of risk assessment under management regulations. Practice is variable.

The overall context has been falling public expenditure and deregulation. Future directions are unclear. Despite the change of government, injuries are increasing, but inspections and prosecutions are falling. There is an absence of a prescriptive and regulatory framework.

Ton Wilthagen outlined the situation in the Netherlands. He described the change in the regulatory paradigm, from: substantive to procedural, problem-solving to prevention, and non-involvement to involvement. He saw risk assessment as either the key method in self-regulation, a paper tiger, or a sham (political economy view). As in the UK, there is a lack of evidence of practice. Features of implementation include: adjustment of primary legislation, mandatory to contract OHS, and mandatory to have OHS services help to produce risk assessment. As for company practice: there is little empirical evidence, less than 30% of small firms have done assessments, quality is allegedly low, and worker involvement is allegedly low.

Regulatory practice of inspectors included use of a variety of strategies: a business as usual, top-down approach; a reserved approach regarding scope and depth of check; a weak connection between shop floor problems and system flaws; drafting of a model risk assessment; off the record advice on OHS management; and limited interaction with works councils. Overall, there may be a period of transition.

Per Langaa Jensen outlined the history of the situation in Denmark, based on legislation since 1972. There is mandatory watchdog provision. He explained the "sidecar" function of health and safety, whereby the inspectors do not attempt to drive the "management motorcycle". Workplace assessment was seen as a means of revitalising practice, with a clear legal basis. Local understanding of risk is fundamental, rather than reliance on outside experts. This can lead to conflict.

Research has been conducted on implementation of these measures. Larger firms are more compliant than SMEs. In general, workplace assessment is seen as successful, providing a basis for discourse. Firms have been happy to pay for OHS when based on sound assessments. The packaging and action plans make it seem businesslike. However, the occupational health services do not see themselves as process consultants. The view is that known problems are being repaired, but forward planning has yet to take full advantage.

The Danish labour inspectorate have an approach comparable with that of **Gunningham** and **Johnstone**, identifying three different classifications.

Alex Karageorgiou described the situation in Greece. The construction sector is most vulnerable. The workers are both supporters and transformers of the structure. The Greek economy has been seeking to meet Maastricht criteria, with 10% unemployment and 5% inflation. Agriculture still employs 20%, with 23% in industry. Among workplaces, 99.4% have fewer than 49 workers, meaning that SMEs are a problem. There is a lack of integration between responsible ministries. Occupational accidents have declined.

The 1989 EC Framework Directive was implemented, building on past provision from 1985 legislation. Employers are obliged to use the services of safety officers and occupational physicians. There is a shortage of skilled staff. Representation rights have been enhanced. Employers have the duty to make risk assessments. Employers have the right to advance consultation. A written assessment is required. There is participative occupational risk assessment, taking a dynamic view. Homogeneous groups of workers are involved.

Theme IV Comments – Laurent Vogel

Laurent Vogel commented on the theme. He described his involvement in the trade union internal debate on OHS management and the Geneva debate. Norwegian trade unions were cautious about the strategy of internal control. The European debate included many of the key questions.

1. One strategy was the Americanisation of health and safety: supported by employers and the UK government, allowing the companies to self-regulate. When discussing working time and youth working, the UK government emphasis was on company autonomy. Multinational or even national companies achieved extraterritoriality under the UK favoured solution.

2. Many other governments and unions wanted to specify minimum standards for all, with risk assessment as a tool to help choose the most efficient and effective means.

Laurent Vogel outlined the context of workplace change, present OHS models and political pressures for effective regulation. The strategy can be chameleon-like, talking about profitability with the goal of furthering human life and safety.

What are we managing, when we manage health? It is not a product of the company. Talk of management implies a top-down approach, prescribing what constitutes health. We know this does not work, because the workers are seen as the subjects of the process. Trade unions see it as important to have a two track approach, with separate priorities understood on the part of employers and unions. This means a social debate. Management is important, but the key issue is to organise the dialogue of the social partners, and to proceed on this basis.

There has been discussion of SMEs and complex production structures. Intervention means qualitative transformation, from management to politics. The same companies can be leaders in one country and laggards in another. We are dealing with the politics of health and safety at a society level. Public regulation is to defend current and potential workers. Self-regulation risks systematic selection of the workforce, with long-term consequences.

He concluded with reflections on professionals. In companies with strong policies of internal management and strategy integration, we see the subordination of experts under the general objectives of the company, and their effectiveness in terms of health is reduced. He cited the example of colour distinction: quality could suffer if colours were confused, but this was not a matter of health and safety. From the company perspective, the test was required as a matter of policy. What will be the situation of preventive services?

V. Assessing the Effectiveness of a Voluntary and Regulated OHSM

Theo Nichols and Eric Tucker
OHS Management in Comparative Perspective: UK and Canada Compared

A political economy approach was adopted, assuming that production is for profit, and not for safety. Workers' interests are often in conflict with those of employers, and there is inequality of power. Societies differ in their levels of inequality, and in their institutional structures. There is a spectrum from USA to Scandinavia.

Eric Tucker described the situation in Ontario, with state support for improved practice and rights for workers, through an internal responsibility system. This sought movement of firms to continuous ethical improvement. Workers argued for more control, corresponding to their risks. Governments and employers were defensive. There was then a concern by employers to reclaim the high ground, using health and safety management to enhance legal defences. Government argued that safety pays, avoiding power struggles. They argued that safe workplaces meant sound business. Evaluation is difficult to date. Labour sees the OHSM system as undermining its interests.

Theo Nichols dealt with continuity and self-regulation in the UK. His focus was on OHS management, and Health and Safety Guidance 65, published in 1991. DuPont were among the consultants quoted. The Human Factors Group investigated the Nuclear Industry, with outdated American psychology. Safety culture is seen as a form of self-regulation. Managements with objectives and clear strategies will do the job better. However, OHS management may conflict with other strategies, which have had priority. Intensification of labour may be to the detriment of quality health and safety. Steel companies using OHS have placed staff under pressure if absent. Reporting was lax. The HSG65 philosophy was to reward good practice: what gets rewarded gets done. Accident reporting has been suppressed, and contract workers face greater pressure. Privatised companies such as in the coal industry are not accountable to Parliament. HSG made no provision for trade unions, drawing on advice from DuPont, and the document made no mention of trade unions.

Team working in a Canadian-owned UK steel company included de-recognition of trade unions. Management appointed safety advisers replaced union representatives. The assumption was that accidents are caused by individual human errors. "Empowerment" does not seem to be in the interest of the workforce.

Richard Wokutch
The Role of Occupational Health and Safety Management in the United States and Japan: the DuPont Model and the Toyota Model

Kaj Frick had suggested considering the DuPont and Toyota models. Since 1960 the relative reported injury and illness rates for Japan and the USA have reversed. There are some problems with the data, so **Richard Wokutch** undertook detailed research. In manufacturing US injuries were much higher. Comparisons were made between US and Japanese plants owned by the same companies. It appeared that there were 6.5 times more non-fatal to fatal injuries in USA as opposed to Japan. Japanese workers saw injuries as a sign of loyalty. Interviews with Dutch DuPont workers are planned. The level of DuPont injuries and illnesses seems very low.

How did Toyota and DuPont do it? They link OHS with productivity and TQM, in line with Japanese manufacturing approaches. OHS is integrated into production and planning, with a mix of behavioural and engineering approaches, an environment of labour–management cooperation, and a framework of regulatory relations and a safety culture.

Weaknesses of the Toyota approach include: cooperative unions and regulators, subcontractor problems, operations in developing countries, *Karoshi* (death from excessive work), under-reporting of injuries and illnesses, and the safety and health implications of lean production. DuPont weaknesses include an absence of cooperation with unions, under-reporting, regulatory relations, an over-emphasis on behavioural approaches. More research is needed.

Peter Westerholm and Rick Fortuin
Modern Occupational Health Systems as Agents of Change

The role of change agents does not always come naturally, but there are some useful examples. The discussion concerns the full range of professionals, and is built more on experience than on empirical material. It acknowledges the additional challenges through new work organisation, in addition to traditional problems. The economy must be considered. Who is the client? Power relations are important. What benefits can be derived? We have to reflect on relations with the full range of stakeholders.

Agendas vary. Health professionals and customers have different agendas, and deploy varying strategies. **Bjorn Gustavsen** has described the roles of external agents, depending on knowledge and responsibility. There are issues of prior experience by the organisation and the change agent, of work organisation, health matters and local relationships. In addition, financial arrangements are important, and carry ethical implications.

Four questions were raised:

• Is it ethical to offer services in which you are not competent?

• Is it ethical to offer services which are not needed?

• Is it ethical to offer services in which you do not believe?

• Is it ethical to offer services with the intention of preserving your own power base?

Rick Fortuin took up these questions, based on experience in Holland. The profit motive is complicating life in the occupational health services. This challenges accounts of the agendas of the client and professional.

David Walters and Laurent Vogel
Risk Assessment and Workers Participation in Health and Safety: a European Overview

David Walters recalled the points made by **Laurent Vogel** regarding different logics followed by employers and workforce. Given the significance of assessing the effectiveness of participation, it is surprising how little information we have. Worker representatives can channel expertise into the practice and process of risk assessment.

What is meant by participation, both representative and direct? There is more evidence on representative participation. This is important for risk assessment, where even less information is available. Regulation, management commitment, and strong workplace organisation were all important.

Changes in work organisation are impacting on trade unions and representation. Trade unions help establish baselines, so their reduced presence can have a multiplicative effect.

Strategies of trade unions have to address this challenge, incorporating non union members. There are numerous consultative mechanisms available, but worker centred approaches are not widely evident at workplace level. Trade union education and training can give an indication of levels of awareness and activity. Training on risk assessment is being redefined. Trade unions are seeking new identities and roles, and health and safety is one suggested answer to the crisis of representation.

Theme V Comments – Claire Mayhew

Claire Mayhew drew on the work on DuPont and Toyota, discussed by **Richard Wokutch**. OHS was identified as part of QM etc. in the US and Japan. Much seemed familiar, given research in Australia. The recent research on casual labour in an Australian fast food company suggests that the DuPont model has been bought and transported. The ideologies have been moved between industries and countries. There has been no evidence of OHSM systems for contingent workers. She reflected on the roles of OHSM professionals with contingent workers, and could not think of cases in her experience where workers would go to the company medical officer, as their jobs might be at risk. This makes her suspicious of the accuracy of OHSM systems wherever contingent labour is involved.

As for worker representation, **Richard Wokutch** had said little of relevance to dispersed disorganised workers. They only participate in management systems if forced to do so. It is a matter of economic pressure. Working conditions are the price of the job. Contingent workers do not participate in risk assessment. They need to be empowered. It is not clear that OHSM can work with contingent workers. Changes in the labour market are happening which result in much poorer OHS outcomes. Inspectors are under pressure, with increased demands.

Evaluation needs to consider a range of factors and ideologies, including the ideological underpinnings of management systems. Does the introduction of such systems enhance managerial power, and diminish the focus on hazards and risks? The evaluators of OHS need to be independent, with the findings kept independent. There is an underpinning of neo-liberalism behind OHSM, and when the mask slips this becomes apparent.

Contingent labour is a case on its own, of growing proportions. Economic survival is the priority. Australian demolition workers contract for one-off jobs, dealing with asbestos at night. They have few formal skills, and competition for work is intense.

Final Session

Nicholas Ashford considered focal issues for social engineers: production systems, and their influence on people who work (and their safety and health), and the modern industrial environment. Schumpeter saw economic growth in terms of technological innovation, which affects work, production and the environment. In a globalised economy, activity concerns increasing trade, and new realities of cultures and standards. Trade is a driving force. Developed nations can sell their surplus overseas, ignoring technological innovation, and then falling behind. We see a race downwards in terms of standards, and pressure to externalise costs such as occupational health and safety. More is now spent on automobile repairs than on purchase: it would be preferable for the car to last without repairs.

Advocates of OHMS want to manage the link between production and workers, but the environment is rapidly changing. The issue is the management of technology. OHMS starts with hazard identification and moves to risk assessment; secondary prevention rather than redesigning processes. This involves professionals and unions, on the basis of a right to know. He argued for a solution focus rather than a problem focus. OHMS approaches can bring companies forward, but the real issue is the management of change in technology and work organisation. This means technology options analysis, looking at alternative products and processes, looking at comparative risk assessment models. We need cleaner, safer production: this is primary prevention, involving a different community of professionals who have not to date been involved in the debate. This involves workers and unions in a right to act, and technology bargaining.

Kathleen Rest found the discussion refreshing, starting with shared assumptions, and looking at how to proceed. This was in advance of debates in the USA. We need to be clearer about what we are seeking: not just reductions in reported injuries, but humane workplaces where workers are valued.

The SALTSA International Research Programme

Anders Schaerstrom provided an overview of the SALTSA research programme in general, and the Work Environment programme in particular, where he is secretary. **Kaj Frick** explained that **Per Langaa Jensen, David Walters** and **Ton Wilthagen** are leading projects supported by SALTSA, dealing with the Web site, SMEs and national policies respectively.

List of Participants

Nicholas Ashford, MIT, USA
Peter Dorman, Evergreen State College, USA
Tom Dwyer, State University of Sao Paulo, Brazil
Joan Eakin, University of Toronto, Canada
Richard Ennals, Kingston University, UK
Rick Fortuin, National Occupational Health Service, Netherlands
Kaj Frick, NIWL, Sweden
Neil Gunningham, Australian National University
Richard Johnstone, Melbourne University, Australia
Annette Kamp, National Occupational Health Service, Denmark
Alex Karageorgiou, Centre for Prevention of Occupational Hazards, Greece
Karl Kuhn, Federal Institute for Safety and Health, Dortmund, Germany
Felicity Lamm, University of Auckland, New Zealand
Per Langaa Jensen, Technical University of Denmark
Tore Larsson, SAMU, Sweden
Kees Le Blansch, Netherlands
Hans Jorgen Limborg, Denmark
Preben Lindoe, Rogaland Research, Norway
Claire Mayhew, National Occupational Health and Safety Commission, Australia

Carolyn Needleman, Bryn Mawr College, USA
Theo Nichols, Bristol University, UK
Klaus T. Nielsen, Roskilde University, Denmark
Pascal Paoli, Ireland (Dublin Foundation)
Michael Quinlan, University of New South Wales Australia
Kathleen Rest, University of Massachusetts Medical School, USA
Andrea Shaw, Consultant, Australia
Lena Skiöld, NIWL, Sweden
Eric Tucker, York University, Canada
Laurent Vogel, ETUC, Brussels
David Walters, South Bank University, UK
Peter Westerholm, NIWL, Sweden
Ton Wilthagen, University of Amsterdam, Netherlands
Richard Wokutch, Virginia Polytechnic, USA
John Wren, University of Auckland, New Zealand
Gerard Zwetsloot, NIA'TNO Netherlands

Reflections on the Workshop

Held at the Trade Union museum, this workshop was very concerned with the practicalities of the workplace, and included a vigorous trade union input.

The context of debate was global rather than European. Australia and New Zealand had provided excellent research partners for the Swedish organisers, and brought a wealth of valuable comparative data and experience. There were insightful comments from participants from the United States and Canada, and from Brazil. The workshop built on pre-existing collaboration, and will lead to further publications and research, including through the SALTSA programme.

The workshop was longer than others in the series, and benefited from joint working of participants in advance, and the circulation of papers. Over the four days many of the themes of the Work Life 2000 series were discussed.

5. Developing Work and Quality Improvement Strategies

The workshop was led by Jörgen Eklund and Bo Bergman of the University of Linköping, and was held in Brussels, 19–21 November 1998, at the Office of the Swedish Trade Unions.

Abstract

The driving forces to creation of good work have changed over different time periods. The unemployment rate seems particularly influential in this respect. For Sweden, but also for many other countries, it is no longer the ideals of socio-technology that push the development of good work. Among other aspects, internationalisation, increased competition and EU regulations have increased the pressure on organisational change and renewal in order to improve quality, productivity and efficiency of operations. This pressure is to an increasing extent transferred to sub-deliverers, i.e. small and medium sized companies. Quality strategies, such as Total Quality Management, are gradually gaining increased acceptance in many companies. Since such quality strategies contain certain humanistic aspects on the design of work, they are potential driving forces towards good work. However, other aspects and applications of quality strategies are not entirely positive, and can even introduce risks and negative aspects for the development of the workers and their work.

The aim of the workshop was to identify interactions between quality strategies such as TQM and developing work, and to identify potential improvements of strategies and methods used in this field, especially for small and medium sized companies. In addition it identified areas of mutual interest or contradiction between the approaches of TQM and or ergonomics (human factors).

Introduction

Bo Bergman introduced the workshop, and anticipated a convergence of ideas from different traditions and disciplines. The focus of attention was real working life, and not simply research. **Jörgen Eklund** set out the practical arrangements.

Bo Bergman explained that the workshop would commence with accounts of the state of the art, and consider how the work might be harmonised to jointly improve quality strategies. The role of research would be discussed, together with the advice that might be given to decision-makers. In response to **Kostas Dervitsiotis**, he indicated that the discussion need not be limited to Total Quality Management. Papers had been circulated in advance. Many of the participants will be involved in the forthcoming conference at Linköping on TQM and Human Factors, and the workshop gave an opportunity for the review of abstracts submitted for that conference. Workshop participants introduced themselves, with brief accounts of their backgrounds.

Yashio Kondo: The Contribution of Quality

The opening keynote presentation was from Professor **Yashio Kondo**, formerly of Kyoto University, and a leading figure in the Japanese Quality movement. His talk reviewed the contribution of Quality in past decades, and looked forward to the next century. He gave an account of the conclusions reached by Quality gurus Juran and Feigenbaum, drawing on experience in particular in Japan and the United States. Speaking at the time of President Clinton's visit to Japan, **Professor Kondo** noted that Japan has been seen as resistant to imports and in need of internationalisation, although it was generally acknowledged that Quality played a central role in Japanese business culture.

Since the end of the Cold War, he argued, the movement towards free market economics has accelerated, and three main economic blocs have emerged: Europe, the United States and Asia. Quality can be seen as the key to competitiveness in the global economy. He declared that competition and cooperation are two sides of the same relationship. Japan has to take an international perspective if prosperity is to be maintained. Core lessons learned by American companies apply also to Japan. Despite the turmoil in national and international markets, it is vital to maintain attention on consistent themes. Among these, he regarded quality as more human in orientation than cost or productivity, though developments in the three areas are linked. Whereas we can talk of a quality culture, we do not sensibly talk of a cost culture or productivity culture. Quality has been a concern for a million years; cost for ten thousand years, since the invention of money; and productivity for a mere two centuries, since industrialisation.

With a primary emphasis on manufacturing, **Professor Kondo** considered different perspectives on quality. He noted that improving quality can lead to lower costs and higher productivity, depending on the creativity of the innovation involved. On the other hand, the creative reduction of cost tends not to improve quality. Manufacturers tend to guess the needs of their customers, and then assess the level of satisfaction with the product provided. The customer may demand quality which the manufacturer regards as excessive. The key is to provide attractive quality, to meet or exceed the customer's expectations. As quality increases, the relative importance of money diminishes, both as product price and as the prime motivation for working. As internationalisation increases, he saw quality as continuing to be a dominant factor.

He closed by quoting "Let's Get Rid of Management": "People don't want to be managed. They want to be led. Whoever heard of a world manager? World leader, yes. Educational leader. Political leader. Religious leader. Scout leader. Community leader. Labour leader. Business leader. They lead. They don't manage. The carrot always wins over the stick. Ask your horse. You can lead your horse to water, but you can't manage him to drink. If you want to manage somebody, manage yourself. Do that well and you'll be ready to stop managing. And start leading."

Discussion

Bo Bergman started discussion of competition and cooperation. **Professor Kondo** saw competition as a great source of motivation: it could not be stopped, but it should be fair. A competitive team built its success on internal cooperation.

Kostas Dervitsiotis challenged the idea of the 21st century as the century of quality, and noted that Japanese banking and agriculture had been less inclined to change and improve. Quality is an issue for total systems, with impact across the culture and the environment. **Bo Bergman** reported enthusiasm from the Swedish public sector, where attitudes are changing.

Eamonn Murphy drew the parallel with Chrysler car fenders, made of rubber and metal, which revert to their normal shape after low velocity impact damage. He saw many quality initiatives as being cast aside after a short period when crisis strikes. The Japanese quality movement had been sustained through a long period of economic success: will the concepts survive the current shocks?

Professor Kondo responded that it had taken the United States fifteen years to recover from Vietnam. Deming's book *Out of the Crisis* had appeared in 1983, and the Japanese view is that if the Americans could recover, so can they.

Colin Drury: Service Industries, Quality and Human Factors

Colin Drury is from State University of New York at Buffalo. He observed that the services sector is growing rapidly, and that quality is of central importance, increasing the need for Human Factors data. In most modern economies the service sector is the largest, driving the development of the post-industrial economy, changing rapidly, making extensive use of Information Technology, and operating on a global scale. He cited telephone call centres, which can be located anywhere in the world; BAA are now managing a number of American airports; cheque reconciliation is conducted offshore, and remote software development is becoming commonplace.

Service industries face ever-increasing competition, obliging them to address the concerns of their suppliers and their customers. There is a constant drive to reduce direct costs, which can result in jobs moving, organisations downsizing, and new locations being developed. Job movements mean for developed countries a loss of jobs, downward pressure on wages and overheads, and a reduction of government controls. For developing countries it can mean more work available, but pressure to expand rapidly, and to develop regional specialisms. This could be regarded as a good development, with a fairer distribution of employment, enhanced competition, greater diversity and the emergence of new specialisms. On the other hand, there can be damaging impacts for the environment, exploitation of workers, the reduction of public services for short-term gain, and the shift of the balance of power towards companies and away from governments.

Service industries deal in intangible goods, with physical items as a secondary product. The service constitutes a change in the state of the customer. Examples given were doctors, banks, car washes, transport and telephones. War could be regarded as a service industry: there are no positive products but the state of individuals is changed.

Services are not wholly separate from manufacturing, but primarily involve the customer, while production can be at a distance. Much manufacturing and agriculture is service, often using indirect labour. Services can be compared in terms of labour intensity, customer interaction and customisation. Electric supply is low on all three. Stockbroking is high on customisation. Repair service is high on labour, and retail conventionally high on customer interaction. At the low levels we can think of service factories, while at high levels services are unique, in terms of service, quality and human factors.

In this context **Colin Drury** defined quality as meeting or exceeding customer expectations. He considered designing products, processes and organisations for quality.

Total Quality Management was not the first approach to quality. He was concerned that quality systems alone are not enough. We must distinguish rhetoric from reality, and recall that research has lagged behind practice. There have been some bizarre unintended consequences of TQM initiatives, including falls in company share values.

He then considered the Human Factors revolution, with the goal of error-free manufacturing and services, also injury-free and conducive to a high quality of working life. User oriented product design is a recent development.

Service, Quality and Human Factors can be seen as interacting. Performance observation can be in terms of temporal quality and behavioural quality, but this can place workers under undue pressure. Service quality can be seen in terms of customer satisfaction; ergonomics can be located within TQM; TQM can be regarded as a subset of socio-technical systems. Perhaps most interesting is the current development in service industries whereby the customer and the operator are considered together: often the role of the operator, or service agent, is being taken on by computers. Examples were taken from banking, investment services and travel agents. There can be problems, as the task may not be well defined, and the customer may not understand the structures and institutions concerned. The agent's task is to find the customer's intentions, and match these intentions to the database. The restrictions may not be well defined, and may not be firm. It is not clear that we can program computers to interpret customer intentions, provide flexible trade-offs between the different restrictions in areas of complexity, understand when and how to go beyond the official restrictions, and provide appropriate managerial backup when rules are bent. As a consequence of inadequate systems, customers may prefer to be served elsewhere, with consequences for service industries. Jobs can be lost, changed, or made more remote from customers, with possible cuts in wages.

The obvious conclusion is that Human Factors and Quality should work together. Ergonomics can learn about strategy, leadership and the management of change. Quality can learn about measurement, automation and socio-technical design. The well-being of the workforce should not just be assumed but measured, taking into account human values. The result should be a reduction in errors, an improvement in service quality and improvement in the quality of working life.

Discussion

There was debate on whether financially strong companies entered for the Baldridge Award, or whether it was quality performance that produced the financial success. **Eamonn Murphy** argued that the service component of manufacturing companies offered the greatest scope for improvement. Many companies are better at responding to technical messages than to messages regarding human factors.

A second issue was whether successful companies have happy employees. **Colin Drury** cited exceptions in the steel and auto parts industries, but there is an absence of research, although it is clear that worker involvement helps.

Bo Bergman: Why is Total Quality Management so Often Unsuccessful?

Bo Bergman located the quality debate in the history of ideas, contrasting the view that "there is one approach to society" with the alternative that "there are no ultimate truths". Taylor had put forward an optimal system, and the alternative is a more open approach.

As a focus for discussion, **Bo Bergman** presented in tabular form a contrast between the paradigms of TQM and traditional models of the firm, with a focus on maximising shareholder wealth. TQM is dynamic and favours self-management, while the economic model of the firm is static and emphasises control. Once the idea of stakeholding was accepted by the firm, the incompatibility was potentially resolved: recent quality awards criteria have included stakeholding. The starting point for a company is usually more than simply the objective of making money, but in order to continue it will need both innovation and quality control. Quality may include the identification of new markets, and creativity entails encouraging diversity.

Eamonn Murphy introduced discussion of the Balanced Score Card approach. **Pascale Carayon** noted that Drucker concluded that there is "no one right organisation", "no one best way", but there can be general principles and values. **Brian Joiner** observed that religions cannot be proved correct: the question in business is what works. **Bo Bergman** responded that if value systems fit the organisation, they work. **Stefan Agurén** described experience with sociotechnical systems with Volvo at Uddevala, and with Time Based Management. Economic models need to be combined with social models, and success requires the consideration of new stakeholders. There could be no single nationwide perspective on work organisation. **John Wilson** argued that socio-technical systems do not really offer a model which can be used to provide operational guidance: there is a middle ground made up of hopes and expectations, but an absence of solid material. In an open society, concluded **Bo Bergman**, there is no single model.

Johnny Lindström: Reflections on History and Core Values

Johnny Lindström is Director of the Swedish Institute for Quality. He argued that historically quality was in the hands of the craftsman, as a dimension of his skill. The process of industrialisation led to a Tayloristic approach, and the introduction of inspections. Work was for money, and control was in the hands of managers. The next step was for delegation to be developed, with operations conducted by small units responsible for self-inspection. Next came a phase of coaching, with ISO 9000, management by objectives and the attribution of personal responsibility. More recently there has been a concern for leadership, in a context of TQM, customer focus, responsibility, corporate mission and flatter organisations.

Structurally problems have developed, as vertical patterns of communication and responsibility have encountered production processes based on horizontal processes. At this stage communication is increasingly horizontal, at all levels, internal and external, and using the new technologies of the Internet and Intranets. With the increased demand for knowledge based decisions, a holistic view is required, involving management by values. The Balanced Score Card approach fits this requirement.

Johnny Lindström then set out the Core Values as listed on the Swedish Institute for Quality card, first produced in 1991, and implemented in companies, hospitals and schools across Sweden. Each was expressed in a phrase, a sentence of definition, and a sentence suggesting particular points to think about:

1. Customer orientation
2. Committed leadership
3. Participation by everyone
4. Competence development
5. Long-range perspective
6. Public responsibility
7. Process orientation
8. Prevention
9. Continuous improvement
10. Learning from others
11. Faster response
12. Management by facts
13. Partnership

One challenge for Work Life 2000 might be to review and update the list.

Discussion

Bo Bergman thought that an essential element is trust. **Colin Drury** asked about measuring the effectiveness of these values. **Johnny Lindström** reported on implementation at village level, and the exposure of the values to a wider audience. **Pascale Carayon** asked if the values were culturally limited: around the table they received support from Europe, the United States and Japan. **Johnny Lindström** emphasised that much had been learned from others, including the major award organisations. **John Wilson**, as devil's advocate, asked if companies might only take part if they regarded themselves as successful, and he queried the real strength of participation.

Pascale Carayon: Balanced Work System and Participation

Pascale Carayon drew on research experience in the United States and France, with particular emphasis on the public sector and community applications. As an ergonomist and industrial engineer, she started by tracing the history of the disciplines. Ergonomics has moved in successive decades since the 1950s, from military, to industrial, consumer, Human–Computer Interaction, Cognitive and Organisational, to the new specialism of eco-ergonomics, for the new Millennium. She then set out the framework for community ergonomics, the study of organisations in a community, and their interactions. It is assumed that priorities are the quality of products, processes and working life, that work systems are important, and that system design can influence outcomes.

A key concept was that of balance: the objective that the overall work system should be improved, or at least not degraded. This involved considering positive and negative aspects, rather than just striving for perfection.

She took the example of the City of Madison, Wisconsin, USA, which has conducted a Quality Improvement Project since 1983, and has produced apparently positive outcomes in terms of public perceptions and worker participation. However, work-loads have increased, and there is a lack of clarity about job duties that can lead to conflict. This raises questions about the process of organisational change, and the reality of empowerment. The process takes time, and is accomplished in steps. One key transition is from reliance on external experts to the confident use of local skills.

A second example was from Ludres-Fléville, near Nancy in France. Here 200 SMEs have been brought together by an association concerned to help with Quality Improvement, based on a model of active participation by companies. A similar approach, at levels larger than single companies, is being taken in poor neighbour-hoods and with minority students. Thus, in a practical way, we bring together the quality of working life, the quality of life, balance and participation. This is partici-patory ergonomics.

Discussion

Kostas Dervitsiotis asked about extra time required for training, and saw this as a matter of overcoming initial friction. There could be no quick fix. **Pascale Carayon** noted that the work in Madison had been running for 12 years before the research began.

John Wilson: Searching for Coherence: Interactions as the Focus for Human Centred Systems

John Wilson leads the Institute for Occupational Ergonomics at Nottingham University. He explored the nature of belief in this area of apparent convergence. Is TQM to be seen as a religion, a philosophy, an approach or a set of tools? Is it, rather, just a slogan? There certainly has been much misunderstanding and misuse of the terminology. SMEs tend to have picked out the pieces they like, sometimes achieving good practice, but rarely sustained, yet they employ a large proportion of the workforce.

John Wilson described recent research in participatory ergonomics, self-directed work teams, work organisation in SMEs, scheduling, information support needs and maintenance, and disturbance analysis. He identified key issues as being strategic leadership (not just identifying a team leader), boundary management, and perfor-mance measurement and reward.

He contrasted two views of modern organisations. In the first, the emphasis is on staff control, responsibility, skill, knowledge and involvement. In the second, the starting point is customer ethos, supply chains, reduced variances, management control and organisational flexibility. He concluded that modern manufacturing and ergonomics can be viewed as either complementary or contradictory, with winners and losers. The direction of change is partly up to management, and partly a consequence of external factors.

He concluded with a new framework diagram linking Human Factors and TQM. Many of the issues debated at the workshop (and indeed in the Work Life 2000 series) could be found and linked. He was left with questions. Does Human Centred Manufacturing need a prior technology framework? Will profit always drive decisions? After all, he declared, managers are human beings. There are no neat problems or solutions. In short: interactions are the basis of a coherent approach.

Discussion

Brian Joiner argued that change is best accomplished by first addressing the big problems in an organisation. **Eamonn Murphy** mused on the differences between TQM and ergonomics, and concluded that the principles are similar but the roots are different.

Stefan Agurén launched the final discussion of the day, by asking if we need managers, given that we are discussing leadership roles for managers and workers. **John Wilson** responded by asking if we need factories, at least as we know them today. Production managers and operators could be the same people. **Johnny Lindström** asked if we need organisations. **John Wilson** concluded that structure is important, rather than current form, paralleling the importance of the process of change as opposed to the content. Perhaps, said **Johnny Lindström**, the job of the manager is to work out structures. **Colin Drury** reflected on the imposition of structures through enterprise resource management systems such as SAP, and registered concern at the consequences.

Klaus Zink: a Broader Agenda Based on Stakeholding

Klaus Zink sees quality of work and quality of life as part of quality, incorporating a stakeholder approach. The economic dimension cannot be neglected. Frameworks are changing. There are changes in structure, with a decline in manufacturing and an increase in services. We are on the way to an information society, with implications for the future of work. Globalisation and pressure of time, the diffusion of responsibility and loss of commitment all force us to restructure work. We have to consider teleworking and virtual companies. Global problem-solving round the clock has begun. Time to market is compressed. Anonymous money is managed by paid managers. Values are changing, linked to demographic burdens and new social welfare challenges. There are new demands on HCI. Governments are trying to limit financial burdens. There are technical changes associated with independence from space and time. This means new forms of leadership and management, new models for working hours, and in essence a new society. Large obsolete office buildings pose challenges.

He offered the results of some brainstorming, with recommendations for government:

- Control mechanisms for floating money
- Pay for work programmes rather than unemployment
- Implement new evaluation systems for welfare (GNP is a flawed measure)
- Support organisations and companies improving the quality of the environment
- Working with TQM principles

As for companies and organisations, he advocated:

- Long-term instead of short term orientation
- Holistic assessment concepts, instead of focus on costs and short-term profits

- Changing reward systems (especially for top managers)
- Broader assessment concepts for institutions assessing other organisations

There are messages for science and research:

- Quality pays
- Investing in people leads to profit
- Systematic collection of best practice
- New forms for work and employment, creating new financial models

Discussion

John Wilson raised some inherent contradictions, relating to time frames. Change takes 5–7 years, but managers think in terms of 2–3 years. Klaus Zink argued that there need to be short-term measures within the long-term context. There need to be some early successes to create motivation to continue. John Wilson argued that people tend to get lost. Klaus Zink sees assessment as a means to continuous improvement.

Bo Bergman raised the idea of customer power, and the means by which trade unions can use their resources. P.-O. Bergström described the Swedish LO plan to develop a consumer approach, and increasing collaboration with European partners. He described experience in Volvo and ABB, and the international reputation contradicted by ongoing conventional behaviour by managers. Klaus Zink noted that financial information is not enough: balance sheets and satisfaction indices tell you about the past, as you row the company boat into the future.

Colin Drury asked who are the forces concerned with long-term thinking. How can we exchange ideas? Klaus Zink said the process is no longer under control. Risk-taking has been reduced: those who fail lose their jobs, but do not take risks on behalf of their teams.

Bo Bergman argued that short-termism is self-reinforcing. Brian Joiner believes in the free market system, and sees many investors and corporate leaders as taking a long-term view. However, the pace of change requires rapid decisions by companies, producing high performance from the people concerned. Companies that push their people hard tend to perform well. Rewards are based on performance over time. Truth will out. We do not know what the answer is: it is over-simple to think that decoupling rewards from quarterly performance would produce a desirable transformation. What do we as society want from a company? Investors seek profits. Kostas Dervitsiotis argued that there are other stakeholders to be taken into account. John Wilson talked about damage to roads by large companies, and argued that a view based solely on company responsibility to shareholders cannot be sustained. Brian Joiner responded that we need to review the ground rules for corporations if we want different behaviour.

Jens Dahlgård: Quality Strategy: a Means to Improve Quality of Work

Jens Dahlgård leads a Masters programme in Quality Management at Aarhus. He argued that the need is to develop work through quality improvement strategies. The presentation went from strategy through to implementation, based on experience of the European Quality Award. He argued that everything that is needed is available, but companies focus too much on the business results, at the expense of the people dimensions.

The aim of a quality strategy is to improve "The 4 Ps": people, partnership, processes of work and products. The fifth P is profit, followed by the sixth P, pensions. Companies find it hard to understand and accept that people must come first.

Research at Aarhus has addressed "the trinity of human needs": biological (money), mental (competences) and spiritual (values). Core competencies should meet mental needs: social/emotional (belonging, status and identity) and growth/intellectual. Techniques for successful learning cover reading, writing and creativity. This is to support the actualisation of intellectual capital. Emotional competences include communication and teamworking, dialogue and discussion. Core values include humbleness, honesty, justice, loyalty, respect, trust, dignity, and altruism. These are seen as independent of time, culture and place. In Chinese the character for "profit" means "payment for trust", a view which also underpins the cultures of Japan and Korea.

It is important to understand how motivation and employee satisfaction can link with corporate strategy and mission. Research assessed responses to the excellence model underlying the European Quality Award. Priorities emerged as people, processes and people management. The gap between current practice and the goal of quality excellence has been clarified. Policy and strategy are central. Managers can come to see that improvements in leadership feed through to policy and strategy, which in turn feed through to people management and processes. Improving people management and processes has clear impacts on employee satisfaction. All too often managers agree that measures (such as development opportunities for all employees) are important, but they do not take action. It takes time to change mind sets. It remains difficult to break out of old patterns of working, and an external stimulus can be helpful.

P.-O. Bergström argued that money continues to talk. Jens Dahlgård wants research that demonstrates the role of enablers (i.e. leadership, people management, policy and strategy, resource management and processes), and the links from enablers to employee satisfaction. We need to focus on areas beyond profit, which has tended to dominate.

Jörgen Eklund: What Are the Consequences of Standardisation and Customer Focus?

Jörgen Eklund is interested in the areas of disagreement between the two disciplines at the workshop.

Standardisation has been discussed in the literature. It can be seen as hampering motivation and creativity, or as supporting participation in development, and enhancing creativity. A Swedish case study took 11 companies who had worked with standardisation. There are different views as to what it is, with many national and international standards. There are measurement methods, and approaches to design and procedures. It has been argued that standardisation of human movements is to be avoided, but tolerances are helpful.

So what is the object of standardisation? It can mean better work content and changing levels of variation. The influence can be positive or negative. Can we predict the outcomes? Is it a question of distinguishing between coercive and enabling standardisation. Who makes and decides the contents or applications of the standards? How are the standardisation and implementation performed? Introducing a written document tends to have negative impacts, but participation and continuous improvement are positive: it is a matter of mental models.

Customer focus involves customers as a third stakeholder alongside employers and employees. These are the key stakeholders who must be kept satisfied, but there are others, who should not be dissatisfied. The triangular model was expanded to encompass the concerns of the different actors. Employers want efficiency, employees are concerned with work conditions, and customers want quality. In the centre we find Fordism and BPR close to employers. Different versions of TQM gravitate towards different corners, while socio-technical systems and human relations are closer to employees. Those movements near the periphery will become seen as fads, while the quality movement includes generic elements which relate to the three main stakeholders.

Customer focus can mean job enrichment, communication, decision-making, learning, involvement, productivity and quality. In other cases customer demands can be unreasonable, causing planning difficulties, stress, time pressure, overtime, unsafe working conditions and conflicts. If the customer focus is pushed too far towards the employers' corner, there can be problems.

Brian Joiner asked about standardisation and gave the example of golf. Should certain movements be standardised? **Jörgen Eklund** argued for sound training, but tolerance of diversity among proficient and expert players and workers. However, golf is not wholly comparable to the workplace. The rules of golf are constant, while the work environment requires constant change and adaptation. **Paul Lillrank** sees the similarities as relevant: when he goes to a doctor he expects that standard operations will be performed in a standard way. **P.-O. Bergström** argued that the goals of working life are more complicated, while **Paul Lillrank** argued that the goals are relatively similar. **Stefan Agurén** indicated that robots are better at standardised work than are golfers. **John Wilson** argued that the analogies are mixed: it is a question of sound procedures and the capacity to go beyond procedures where required. We should design work for flexible human postures.

Jörgen Eklund asked for responses to the triangular model. **Stefan Agurén** argued that the model is not new. It should be applied to each process at each level in the organisation, and not used exclusively at the strategic level. **Colin Drury** prefers to see axes rather than corners, and a three-dimensional model.

Eamonn Murphy declared that a company has standardised aspects of its work if it can predict them. Optimisation involves enhancing capability. Innovation typically means technological improvement, for "two from one". The next step is standardisation again. This is a less absolute model of standardisation: the object is predictability, and there is no dispute between the two communities of quality and ergonomics.

Kostas Dervitsiotis spoke of part standardisation, while retaining flexible configuration in other aspects. The example was laptop computers. It is a question of determining the level at which standardisation should apply. You standardise the alphabet, not the final written product.

Paul Lillrank: The Quality Snowman: Quality and Good Work

Paul Lillrank is a social scientist working in an engineering school at Helsinki University. He has kept away from quality and working life for a number of years, objecting to insults to Taylor, whose work he respects. Recently he has returned to the literature, which has the appearance of a ghost town. There has been little about business, amid the structures and social systems. He has now returned to the field.

He offered the "Quality Snowman" model. The small head is quality philosophy. The big stomach is product technology and customer-specific know-how. The feet are quality engineering and managing for quality. Having himself been insulted by quality consultants at his university, he could understand the problems faced by business.

Taking the value chain, he argued that the key is the profitable outcome. Changing mechanisms need to be demonstrably linked to the financial bottom line. Socio-technical systems have rarely addressed this. He argued that net value is the difference between utility and price, at the point of exchange. This discipline of the market economy can sometimes be neglected in what he called the planned economy of Sweden.

Conformance quality was the key to mass production, as an outcome of industrialisation of previously craft processes. Since the Second World War, the intention has been to remove the poison pills, such as boring jobs, products and organisations, without killing the goose that laid the golden egg of middle class prosperity. There has been little innovation in production, while the emphasis has been on making people happy. We can see businesses preferring business process re-engineering and lean production.

He considered the long history of concern for quality, since ancient times, and moving recently from an industrial focus to quality management and then customer focus. The new era is one of non-routine systems and work. In Nordic countries the taxman has in effect solved the problem of automated work, leading to its replacement, at least locally, by robots and the relocation of employment.

Processes can be repetitive or sequential. How can we deal with the quality of working life in these areas of non-routine work? The introduction of quality systems can help to reduce stress in this new work environment. Describing the processes in a way that creates some order can be healthy. Crisis management has become a way of life, and stability is welcome. The other approach is to introduce clearer meaning into work in the business. If people understand the links between their work and the success of the business, this can improve the quality of their working life.

Discussion

Jörgen Eklund wanted to add the element of fun: even if work is busier, it can also be more fun. **Paul Lillrank** wondered whether fun could be managed. It cannot be guaranteed, but a conducive environment can be provided. **Klaus Zink** described a German quality award-winning hotel, offering fun. However, he argued that the socio-technical approach is about more than about making people happy. **Jörgen Eklund** agreed that it is a matter of designing an environment that minimises obstacles to fun.

Klaus Zink asked about the meaning of business. **Paul Lillrank** noted that Finnish SMEs could not relate the concerns of quality professionals to the their bottom line concerns. They need to see what it means.

Kostas Dervitsiotis asked about the poison pills brought by scientific management. Large volumes, centralised control, division of labour, standardised products, slow bureaucracy and boring jobs. The model disregards the environmental costs, and gives misleading conclusions. **Paul Lillrank** argued that customer focus and incorporating the environmental dimension could mean a requirement to remodel TQM. **Brian Joiner** saw the answer as changing the ground rules, not allowing companies to continue to externalise their environmental costs. **John Wilson** spoke of social audit, addressing these issues. **Brian Joiner** spoke of taxing waste and pollution, while cutting taxes on wages.

P.-O. Bergström: From Childhood to Adulthood in the Workplace

P.-O. Bergström spoke as a former steelworker. He talked about blue collar workers, who are still different from white collar workers. They work in the same workplace, without really understanding each other. They have different goals: surviving the shift, for example. The worker stays in the workplace, in the work group, while the manager may travel. There is still silence in the workplace: people say what the other wants to hear, rather than what they think. This is rational, as people fear losing their jobs. There are still two different worlds.

He argued that building an organisation is not hard in itself: the challenge is maintenance and solving problems over time. Workers are now seen as the key resource of the company, but are better seen as human beings. As a steelworker he was treated like a child: it would be better to be seen as an adult; at present we are teenagers. It is a matter of taking on responsibility: this is a feeling that cannot be commanded. With information and the power to make decisions, people feel better. Rules provide the foundations, with shared core values collectively arrived at. Disciplinary matters have long been collective, but need to be based on individual respect. The punch clock is organised to deal with the 5% who will not respond to modern demands. When buying a mountain bike, we do not normally specify that it must have been assembled by workers using punch clocks. All workers are moving towards customer contact. Wages can be seen as maturing: as with children, the early approach was piecework. Wages are still used as the driving force, but a larger proportion of salary is standard. The 10% quality bonus is common. Training is still limited and badly planned. It is seen as a cost, not an investment. The challenge of competence is not being addressed, as it is seen as too difficult. Working hours have become more flexible, but with a framework.

Discussion

Jens Dahlgård observed that McGregor's "Theory X" is still in action. Managers realise that what they are doing is stupid, but they continue. **P.-O. Bergström** sympathised with pressures on managers, including time pressure. **Jens Dahlgård** noted that managers perform differently in their social life. **Paul Lillrank** asked about the origins of treating workers as machine parts or children. What are the ideological or practical roots? **P.-O. Bergström** referred to the legacy of Fordism. **Paul Lillrank** argued that the upper classes have always oppressed the lower classes, peasants and slaves. Things are better than they were. **P.-O. Bergström** described gaps in the knowledge and understanding of students being trained for management careers, such as a lack of experience of unions. There are no quick fixes.

Stefan Agurén: an Employers' Perspective

Stefan Agurén welcomed the initiative behind the workshop series. He speaks from the management side, and there can be worries about initiatives from politicians. He fears far-reaching simplification. Allan Larsson's Green Paper on Work Organisation for the European Commission was launched last year. The workshop shows that there is no one simple process of change. Inside companies we can see a mix of experience, and a journey to the future. We need situational understanding of work organisation.

He described a paper mill near Linköping a decade ago, where they had small customers with limited storage capacity. With rapid shifts between orders, quality standards had to be achieved rapidly. The company had planned a move to new systems, but an explosion occurred in a paper machine, destroying vital parts, and accelerating the technology change, using the same people. Training was essential, and the computer supply company failed to provide what was needed. The problem was that workers saw the threat to their years of skill and experience, which were no longer valued. Two researchers were working on materials resource planning, and interviewed the workers about their work. There were different strategies used by different workers, and through dialogues and process descriptions they identified the proper roles for the computer. The relations between workers and researchers changed. Managers were surprised by the outcome, and offered to take some workers to England to meet customers. Workers developed interest in customer needs, and saw different requirements, going beyond the official manuals. It was clear that the workers regarded machines as if they were alive, carrying forward language from pre-industrial craft work. Workers addressed the challenge of improving absorption characteristics of the paper, as was wanted. In the absence of managers and white collar workers, they conducted experiments and succeeded, demonstrating the benefits from workers taking such initiatives. The change was experimental, and could not be foreseen.

How can we convey the message to politicians from this kind of unique story? How can politicians and trade unions learn from the reality of work organisation? He hoped that researchers will give local parties instruments, models and tools to develop their own organisation to meet their own needs.

Stefan Agurén has learned a great deal from work on the quality of working life. He was surprised that they continued the conventional natural science methods, with little practical benefit. Action research has supported processes in the workplace. Researchers can offer tools, developed from experience and made available for local use. Learning about organisational development must be local.

Discussion

Bo Bergman asked how those present could participate in the process described. **Stefan Agurén** asked about developments in approaches to research in the field of quality and ergonomics. Is it now multidisciplinary or single discipline based? **Jörgen Eklund** regards the workshop as in the multidisciplinary tradition. **Bo Bergman** referred to the organisation of academic work. **Johnny Lindström** left academia, but sees too many structures hindering progress. The narrower the work, the easier is academic promotion. There was discussion of new research funding arrangements in Sweden. He sees a need for more interdisciplinary research from the Swedish Institute for Quality.

Brian Joiner considered interventions which produce results, and pointed to research and learning among consultancies and within corporations. There needs to be cross-sectoral work between academia, corporations and consultancies. **Stefan Agurén** noted roles for consultants in this field. **Eamonn Murphy** recalled action learning, involving executives in reflection on their roles as change agents. **Kostas Dervitsiotis** spoke about action research and futures research. Is a company in a situation where change is desired? There can be contentment, denial, confusion and renewal, envisioned as separate rooms. He finds this active approach as more attractive than simply offering theories.

P.-O. Bergström highlighted key features of the paper company case study, including the presence of the key people, crisis, external intervention, the study visit to the UK and the role of the night shift.

Klaus Zink noted the different backgrounds of the two traditions, bringing different experience to the use of the same language. **Colin Drury** noted that work has to be done in the field, but this does not invalidate conventional research. Academics should not be thrown out. **Kostas Dervitsiotis** favoured developing new roles as learning enablers.

Brian Joiner: Planetary Quality: The Future Ain't What it Used to Be

Brian Joiner has had successful careers as an academic and quality consultant, but is now primarily concerned with sustainability. He looked at the environmental dimension of quality. He highlighted an advertisement raising concern over toxic contents of children's toys: it criticised the US government for blocking restrictions which would threaten growth of toy companies. We can expect more collisions between production and the environment. He outlined the concept of the "funnel", as developed by the Natural Step's Karl-Henrik Robert. Living systems such as those that provide us clean air, water, soil, climate control, fish from the sea, are in decline due to overuse or abuse. At the same time, demand on these systems is rapidly increasing as popular affluence and the use of technology increases, causing the margin for action to be narrowed. Robert and his colleagues evolved four conclusions from basic science.

1.   Matter and energy cannot appear or disappear

2.   Matter and energy tend to spread spontaneously

3.   Biological and economic value is in the concentration and structure of matter

4.   Green cells are essentially the only net producer of concentration and structure

What does this mean for companies and society? From this basic science Robert derived four system conditions:

1.   Substances from the earth's crust must not systematically increase in the biosphere

2.   Substances produced by society must not systematically increase in nature

3. The physical basis for the productivity and diversity of nature must not be systematically deteriorated

4. There needs to be fair and efficient use of resources with respect to meeting human needs

If we do not pay attention to this basic science and the four system conditions, the environment will not be able to provide us with the services on which we depend for survival. These issues are central to quality of production and consumption. The problem is partly due to population increases, but even more due to consumption factors: the richest 20% consume 83%, and cause the bulk of the problem. There is massive waste in the current situation.

The implications for business are stark. Offending toys have been withdrawn from the shelves just before Christmas. There had been industry efforts to contain the situation, but those companies hit the wall. A strategic company starts to make changes well before it hits the wall. Changing a company direction takes time. It is better done strategically rather than reactively. Unfortunately, today, environmental disasters are often seen as increasing the GNP, through the clean up effort required.

Companies such as IKEA, Electrolux, Scandia Hotels, Interface and McDonald's Sweden are examples of companies changing direction. Government subsidies today are still protecting the *status quo* in many industries.

There are many implications for Quality and Quality of Working Life:

1. New welfare challenges, new opportunities

2. Without planetary quality there is no quality of products or working life

3. We need to deepen our understanding of the new context

4. We need to become leaders of needed changes, giving maximum benefit with least impact, and supporting pioneering organisations

We live in interesting times. Will this be a curse or a wonderful opportunity?

Discussion

Paul Lillrank noted that the agenda had been discussed by the Green Movement. What would the world look like if the agenda was followed? How can the political message be formulated? **Brian Joiner** talked of possible futures, and the need for broad involvement in determining the future. **Paul Lillrank** asked if we might be revisiting a past era. **Brian Joiner** recommended that we look forward and take short steps. He has become a vegetarian, and plans to build a sustainable house. **John Wilson** can imagine answers for each of the sub-problems, but the overall effort of will required is enormous. **Brian Joiner** gave a favourable interpretation of America beginning to move on global warming. **John Wilson** noted that UK progress on global warming was only achieved by closing the coal industry. **Bo Bergman** emphasised the role of democracy and judgements of customers. **Brian Joiner** described Eco-Teams, taking action at the household level. **Colin Drury** argued for informing the customer.

Kostas Dervitsiotis: Quality of Work Life

Kostas Dervitsiotis addressed the expected needs for the future, in terms of working life, compared with what will be feasible. There is a gap to be filled. He traced links between the quality of work life, life and the environment. Work life was measured in terms of income, enabling one to lead a good personal life. There are clearly areas of overlap and convergence, so that we need to discuss the quality of life. The need is for sustainable communities. The quality of environment includes natural, cultural and community dimensions.

He provided an historical account of quality of working life, starting with Taylor and what went wrong with his approach. From experts solving problems, we moved to everybody solving problems (participative engagement), then experts improving whole systems (socio-technical systems), involving workers in whole systems improvement, and then (the next step) getting all involved in improvements for sustainable communities. The current phase concerns learning organisations. Sustainability is becoming a critical issue. We are like the smoker who must quit or die of cancer.

He presented the approach of Emery and Trist to human needs, using a TQM style diagram. The satisfiers constitute a negative account of quality of work life, with motivators for positive change: variety, challenge, elbow room for decisions, feedback, learning, mutual support and wholeness etc.

We can thus approach changing work life positively or negatively, or through a balanced mix. Policies include improving communications, challenging roles, investing in training, providing opportunities to make a difference, creating flexibility in employment, and facilitating learning. He argued for using the TQM "house of quality" approach to quality of work life. We should set out motivators, policies for improvement and relative weights. This provides a basis for discussing quality of work life.

He set out three scenarios for the new Millennium, "Business as Usual", "The Big Compromise" and "Sustainable Communities". He considered the needs of different patterns of production and the need to change consumption, in view of environmental constraints. Organisations need to determine their mission, take into account competition, make sure that performance is aligned to external realities, and provide infrastructure support for change.

Discussion

Klaus Zink was worried about the use of the "house of quality" approach to quality of work life. It was likely to produce "one best solution", which would not be a recipe for all. **Kostas Dervitsiotis** noted that he had set out a specific organisational framework for handling relevant elements and interactions of the QWL issue, not a universal panacea. It was framed to be place specific, culture specific and time specific. The approach is also culturally dependent: for example, less weight is given to freedom of choice in oriental cultures. **Jens Dahlgård** noted that there may be varieties of motivators within a single company. **Stefan Agurén** observed that each individual may see things differently, and their views may change over time. He asked whether work life is collective or individual. We tend to see quality of life in individual terms, but work life is considered collectively. **Klaus Zink** recommended starting with the core values, then diverging.

Eamonn Murphy: Quality of Work and Life

How can EU Member States survive on 10 euros per hour minimum wage, a 35 hour working week, 20 days paid leave and full employment? What can TQM contribute? What would the difference be between high and low quality workplaces? There are two key features: personal autonomy and complex social interaction.

From a tradition of adversarial attitudes, we have moved towards autonomy. Quality has been seen as an obstacle, but everybody needs to be made responsible through teamwork. Personal responsibility to family, society and oneself has moved on to an atmosphere of personal freedom, leaving the current conflict between individual rights and the common good. The final key influence is the economy: we can no longer afford supervisors, and the tendency has been to de-layer, resulting in self-directed work teams. Autonomy is an overall objective, which we tend to muddle with different perspectives.

The second focus is complex social interactions. This is based on a systems approach to understanding based on Ackoff's theories. On the horizontal axis, we see mechanical, biological and social models. On the vertical axis we see the defining elements of the system: purpose, function, measurement, characteristic, interaction and environment. The mechanical model has no purpose beyond itself: it is robotic, and measurement is of inputs. The biological model adds an organisational dimension. The social model is more complex, with numerous purposes and no single owner. Love, for example, goes beyond system metrics, and must be inferred. It involves people, rather than organisations.

TQM has ventured into this difficult area. It is not a recipe for solutions, but an attempt to make sense of the social model in a manufacturing environment.

Discussion

Kostas Dervitsiotis asked whether the biological model includes human beings, who have a number of purposes. **Eamonn Murphy** responded by explaining the origins of the model, with a hybrid history. He described himself as looking for his lost intellectual keys under the street lamp, because it was the only place there was light. When we measure companies, inputs and outputs are under the lamp, but the key is still in the dark.

Daniel Vloeberghs: Perspectives on Quality and HRM

The presentation was in three stages. The first focus was on ISO 9000/9001 implementation in Belgium, which was seen as somewhat mechanical. He then dealt with teamwork in the context of Quality Management, and finally he considered the contribution of Human Resource Management to Quality and beyond. The perspective was of the Personnel Department, who had rarely played a leading role. If they want to influence the future, they need to take a more active role.

Turning to teamwork, he compared rhetoric with reality. The work had been conducted with trade union partners, with efforts since the 1970s. The question concerned the readiness of industry partners to deal with new concepts. Management have been a major bottleneck. Teamwork is not synonymous with humanisation and job enlargement. He distinguished lean production and socio-technical approaches. Tayloristic teamwork is also possible, working harder as well as smarter. The young motivated worker can threaten the working life of the older worker.

He set out a new role for HRM in achieving a world class, high quality organisation. HRM is seen as a means of adding value, with different roles as change agent, champion and expert. He offered concrete illustrations of the new kind of partnership agreement, from Digital in Belgium, from the perspectives of the partners, and outlined a proposed implementation process in Europe.

Discussion

P.-O. Bergström asked about recent changes in teamworking in Belgium. Daniel Vloeberghs responded that little has changed in traditional workplaces, and the role of unions is limited in consultancy and new technology areas. Kostas Dervitsiotis asked about the double face of teamwork: was this due to the dominant economic model of business organisations? Daniel Vloeberghs noted that the American model has been dominant in Belgium. It is the shareholder who decides. Klaus Zink noted that 90% of firms in Europe are SMEs, and 50% of the workforce are employed in SMEs. Stefan Agurén reported a major change in the balance in Sweden, with an expansion of institutional ownership of shares. Jens Dahlgård was interested in the practical focus on partnership agreements. Daniel Vloeberghs reported that the process at Digital, now part of Compaq, had generated considerable dynamism. It was conducted worldwide, with different priorities in USA/Canada, Europe and Asia Pacific. Working through focus groups, from descriptors of priorities, they generate essential commitments. Johnny Lindström referred to work on leadership and quality in major corporations, noting that quality is a matter of people, not of techniques.

Report from Theme Discussion

There was a morning of spirited discussion in two groups, bringing together experience from the two previously largely separate traditions of Human Factors/Ergonomics and Quality/TQM, then coming together for presentations and plenary discussion. The intention was to find out whether there was a shared language and a common set of concerns. Participants came from Europe, the United States and Japan, with strong contributions from Swedish employers and trade unions, grounded in workplace experience. There was no formal constraint on the discussion, as the workshop forms part of an ongoing process, leading to the conference in 2001.

The general context was of change, with globalisation and changing working conditions. Europe has to give attention to what we want to preserve. Arguments were developed at numerous levels, including considerations for national and international politicians regarding a balanced continuous improvement approach to addressing major environmental problems as they affect the quality of life and work life. It feels as if the system has let us down. Work is increasingly intensive and pressured. Information Technology advances, and ageing of the workforce, complicate life in Europe. There is a demand for instant results.

The workshop was concerned with quality beyond the requirements of ISO 9000, covering working life, and life in general, including the unemployed. Such issues have been addressed, but not as integrated strategy. The experience of the set of international quality awards has been that understanding of quality has progressively broadened.

The target audience for the workshop starts with Allan Larsson and DG-V of the European Commission, with responsibility for social dialogue in Europe. Through him the ideas can reach a wider political audience and top management. There is an opportunity to influence decisions. Many people have no work, but nobody just works: we remain human beings, with a whole life to live, and roles as consumers. It is a matter of business success in the long run.

Can we link the problems we have identified with the solutions that we can offer? We can offer mental models, ways of thinking and taking decisions. If scientists have developed an understanding of the connections between different areas of technology and work, do they have a responsibility to share this knowledge, and inform decision makers in the political and business arenas, so that progress can be made in addressing the problems? Do scientists from TQM and Human Factors have particular insights? If so, how can these insights be communicated to best effect? How can we adhere to the set of core ethics that achieved general support earlier in the workshop?

There is much from Quality that is relevant to globalisation. Simply working on working life is not enough. The point is that globalisation threatens the Quality of Working Life. This group could have an ongoing function, addressing this agenda. How are we to proceed? There are problems in involving senior managers, and managers in general. New measures such as Balanced Score Card would be helpful, assisting the taking of practical actions. Perhaps there is a need to create crises, or a sense of crisis, as a stimulus to change. We should describe success stories and cases.

What is required from the European Commission? There is a need for the new work life evaluation tool, which has been prepared by the metalworkers, Volvo and others, setting out five levels of work organisation, and tested on the shop floor. It will be available in English and German as well as Swedish. We need some Quality of Total Life indicators, broader than GNP. A European wide customer satisfaction index would help, with information on quality of working life. Tax benefits should be available. Real environmental costs should be met, as with health and unemployment costs.

Universities could improve management and leadership courses, taking on the new agenda. Research is needed on both quality and quality of work life, considering the impacts of short term focus, and positive results. Data should include indices of customer and employee satisfaction, and quality of work life.

At a level closer to the workplace, there was concern to support the development and operation of teams, in a manner that improves the performance in meeting needs. The nature of work has changed, with flexibility, information technology and network organisations seen as particularly important. The roles of managers are changing, with an increasing delegation of responsibility to teams, to enhance the effectiveness of production. Teams in the future may operate as companies do today, meeting the needs of society, recruiting staff and developing strategy for economic, technological and human resources issues. Learning is essential for individuals and the organisation. Those concerned require education, continuing development, and the means of exercising authority. TQM could be an enabling mechanism. Focus on the process, it has been argued, and the workers live.

How can TQM tools help enhance the success of those engaged in this new agenda? Do we expect organisations to continue to operate in the same way? Team members may be drawn from different groups, with less hierarchical patterns and different cultures. Workers may all in essence become white collar workers, with more autonomy and realised maturity. Communication between individuals and groups in supply chains and networks may be considered in terms of quality. It may also be necessary to raise consumer awareness. This has implications for working life, for trade unions, the social dialogue, and for politics. What is new? The ideas of team-work and self-managed teams have been discussed for some time, but now there is less reliance on experts and more self-management.

There is a real sense of urgency. Responding to environmental challenges is not a matter of choice, but has become essential. The quality of life and work life is affected by environmental disasters, which cannot simply be dismissed from consideration. There is an externally determined real world agenda, which includes the remaining workshops in the Work Life 2000 series culminating in the conference in January 2001. There may be messages from researchers in Quality and Human Factors, including case study accounts.

The debate has broadened, from the narrow field of work life, to life in general, with a concern for humanised consumption as well as humanised work. How should we proceed? It was argued that the traditional Swedish approach to designing jobs needs to be replaced by a broader more holistic approach, based on the stakeholder model, associated with economic policy and the interests of consumers. There remain areas of work that defy automation, and must continue, though they are less than ideal for the workers concerned. There is always scope for improvement in work processes and in work life, in the context of producing improved products and profits.

On the other hand, perhaps the more the discussion of Quality is broadened from the core of TQM, the less we can achieve. An intermediate stage may be to envisage a facilitating enabling role for TQM, implemented properly. For this approach to be successful, we need a clearer account of the TQM approach. This brings us back to the nature of multi-disciplinary activities. American TQM has both the benefits and the drawbacks of being "a list of stuff": it is a collection of what works. Given the necessary supporting evidence, American TQM experts will happily incorporate Quality of Work Life elements in the mission to deliver business success. However, the models of TQM in the USA and Europe are not entirely the same. In Europe Quality of Work Life is an early priority, and thus makes the links with Human Factors work. However, the appeal to business has been less than in the USA.

After a period of convergence, as the workshop drew to a close the different tradi-tions re-emerged. As the discussion ended, each expert could be observed subsuming favoured conclusions within their own discipline. Is Quality dead? Is it just a matter of good management? Is there an elegant European theoretical model to match American pragmatism? How do we balance the different elements of the market, the customer, the employee and the environment? The starting point may vary from case to case, but the core values to be included remain the same.

There is no "one best way" towards the achievement of Quality at a global level. In Japan it is not culturally possible to make the workforce unemployed, whereas this presents no inherent problems in the USA. In both cases people are required to work harder than before, threatening their Quality of Life. Within the Quality movement in the USA, we can distinguish different traditions: TQM sees itself as concerned with a particular approach, dealing with flawless processes and cost reduction. The European perspective is different. There are issues of design, participation and Quality of Life. TQM, taken strictly, distances itself from these Quality of Life issues, reverting to processes. It can then be asked whether some companies that profess to follow TQM in fact do so: their processes are often flawed and the business ineffective.

There are questions of individual choice and business policy, with social and political implications. Do people really choose to work as hard and as long as they do? To what extent are employees really stakeholders whose voice is heard? Should the real priority stakeholders be shareholders and customers? In the light of what we now know, can we consider the community and the environment other than as an afterthought? What difference does it make in practice? Given that within Europe we can begin to agree on the set of core values, how should we proceed? How should businesses change? What do they need to prompt them to change?

Bo Bergman summarised the discussion, noting that the different objectives under consideration do not have to be seen as contradictory. Where we encounter contradictions, we have to change the rules by one means or another, for example in the case of the environment, or by enhancing the power of the informed consumer. As European citizens, we need to find ways of going in the same direction. We have assembled fragments that can be put together in a bigger picture.

Brian Joiner compared European with US views on customers. There is more sophistication among customers in Europe. This offers different ways forward. The workshop has brought together interesting people, and we expect valuable results.

Workshop Participants

Prof Yoshio Kondo, formerly Kyoto University, Japan
Prof Colin Drury, State University of New York at Buffalo, USA
Prof Brian Joiner, Joiner Associates, USA
Stefan Agurén, Swedish Employers' Confederation, Sweden
P.-O. Bergström, LO, Sweden
Prof Pascale Carayon, France and USA
Prof Jens Dahlgård, Aarhus University, Denmark
Prof Kostas Dervitsiotis, quality consultant, Greece
Prof Paul Lillrank, Helsinki University, Finland
Dr Johnny Lindström, Swedish Institute for Quality, Sweden
Dr Eamonn Murphy, quality consultant, Ireland
Prof Daniel Vloeberghs, University of Louvain, Belgium
Prof John Wilson, Nottingham University, England
Prof Klaus Zink, consultant, Germany
Jörgen Eklund, Linköping University, Sweden
Bo Bergman, Linköping University, Sweden
Prof Richard Ennals, Kingston University, England
Lena Skiöld, NIWL, Sweden
Maud Werner, NIWL, Sweden

Reflections on the Workshop

The workshop set out to encourage debate between the different traditions of quality management and human factors, and succeeded in increasing mutual understanding. It became apparent that quality management does not mean precisely the same in Japan, Europe and the United States, as the cultural contexts are different. It was possible to construct an overall diagrammatic representation of the field in which different initiatives from the two traditions could be located. The view of the workshop was that an approach to quality management that did not take account of human factors and of the environment would be of limited value.

The workshop cast light on some of the issues surrounding standards, as different traditions were brought together, referring to standards as part of finding a common language.

Work Environment

1. Space Design for Production and Work

The workshop leader was Jesper Steen, of the Royal Institute of Technology Department of Architecture and Town Planning. It was held at the Office of the European Trades Union Congress, Brussels, 20–21 April 1998.

Abstract

In principle, the built environment is a necessary prerequisite for all production and work. Space and material constitute both the means of production and the working environment, and interact in a complex manner with social phenomena.

In the task of designing the built environment, choices of different kinds are made. What possibilities for action there are in this process, and thereby what alternatives appear to be possible, depend on the knowledge of the activity and building in question, as well as the choice of perspective.

At a basic level the built environment contains both practical and social issues. Buildings must be organised so that activities are able to be carried out in a functional manner. At the same time we know there are several different solutions which can function for a certain activity. The social issue deals with how space is organised, how boundaries and contexts are created, and thereby how the built environment interacts with the order which is created in other ways in an organisation and in the community. What options are perceived, and which of these are selected, depends on the insight and sense of priorities on the parts of the designer and the decision-maker.

Both the practical and the social points of view, furthermore, have a rhetorical basis. How a building functions may be more or less clearly expressed in its architecture. And the built environment is full of signs relating how individuals, groups and conditions are valued.

In the early industrial architecture it was usual that one sought solutions which both functionally well suited, and clearly expressed, the specific activity. Changes both connected with new production methods and expansion had the effect that tailor-made solutions became regarded as too restrictive. Instead one attempted to find more general solutions for buildings, a mode of construction more independent

123

from the actual activity, and where the buildings were reduced to a minimum. On the other hand, changes in current work organisation with increased decentralisation have revealed today that large industrial buildings are not functional when "factories within the factory" are desired. To divorce construction from the activity does not solve all problems.

Therefore it is interesting to better understand the way in which the built environment is able to interact with, and provide scope for action for, the activity. Are there certain qualities in the built environment which are more general than others, which architectural solutions create robust conditions able to support changes? Apart from what is possible in terms of building technology, it is also necessary to distinguish the mechanisms in the running and management of real estate which influence the sphere of action of an activity in spatial terms.

This means that the need for functioning processes of communication, on the one hand, between various groups and sections of the activity, and on the other hand, between representatives for the activity and those responsible for the construction and management of buildings, becomes obvious. It is necessary to develop language, terms and images which facilitate these processes, and make the design work a collective process. Experience also shows that the built environment is able to constitute an effective point of departure when discussing more diffuse issues concerning content and organisation in processes of change.

In the workshop, the above issues were discussed using case studies which include car assembly, the food processing industry, and office activities. In addition, the connection between individual production units and the urban environment, and between business development, town structure and existing buildings were taken up using examples from older industrial zones and environments for small firms.

Keynote Speeches

Participants had prepared papers in response to the initial call, and had exchanged drafts via the Internet, meaning that each had a chance of reading the set of papers before the workshop began.

Thomas A. Marcus: Space Design for Production and Work

Thomas Marcus of Strathclyde University focused on rhetoric (representation in language of an ideological mental image), the real world (representation in ordinary language), and key questions for research. He had read and analysed the set of papers, seeking to identify key research areas and common language. His presentation was short and provocative, challenging the language which we routinely use. In general, he saw the papers as full of adolescent optimism. He satirised organisations as expressing bland commitments to the obvious. It was not worth claiming proudly to be in favour of quality and excellence: who was going to announce an aspiration to shoddiness or mediocrity? He was scathing in his criticism of empty pronouncements about learning in the workplace. He argued that unless we could operationalise the nature of that learning, and evaluate whether learning had increased after a particular set of events or interventions, such talk was redundant. He warned that it is harder to use the relevant research tools than is usually realised.

Turid Horgen: Excellence by Design

Turid Horden gave a presentation of work at MIT, noting that the nature of work is changing while many aspects of the environment stay the same. Many organisations are seeking to produce the same products as before, but faster, better and cheaper. Space becomes an artefact to be used sparingly.

Her focus was on innovation strategies, where organisations regroup and redefine their activities. Design can embody different views of work, with functional separation and structures of control, and she highlighted the organisational counterparts to spatial arrangements. She illustrated the Space, Organisation, Finance and Technology dimensions of work practice, where change in one dimension affects all dimensions, and discussed the combination of game theory with design theory to form a new design game. She offered brief analyses of different projects, ranging from integrated to Balkanised. Design is, she said, a political act which helps us to learn about ourselves. She saw new design as happening between the disciplines.

Degenhard Sommer: The Planning of Industrial Construction – From the View of the Practice-Accompanying Science

Degenhard Sommer, speaking from long academic and practitioner experience, based his presentation on the design and construction of a new development centre for Mercedes Benz. He highlighted issues of complexity that raised questions of control over the planning process: the contractor must not be allowed to dominate. Old design processes are too slow, producing work that is already obsolete on completion, and he argued the case for transitional and partial solutions. The process is complex and dynamic, but produces static buildings as outcomes.

He saw the world as a process, and recommended a holistic process approach to design. The system will include elements of chaos, as paradigms undergo change, and interdisciplinary perspectives are considered in a decentralised approach. Planners need to be integrated into the network, in a form of simultaneous engineering. Groups must be able to work together or via networking. He noted the pace of change in information systems over the life of a project, and indicated that responsibilities must be seen as maintained throughout. Integral planning should include infrastructure, considered at a higher strategic level than facilities management. Central and decentralised areas, main and sub-functions should be differentiated. The results of the process are not products, but systems. This work requires dynamic project regulation and effective project leadership. The degree of complexity requires partners in new virtual organisations, working under dynamic contracts, led by process companions, capable of leading projects whose end results are not known.

Shorter Presentations of Papers

Ritva Lappalmainen: The Office of the Future: Where Are You?

Ritva Lappalmainen has a background as a designer in Finland, and works as a consultant in office design. She took the core model of a table, with a person on top sustained in a level position by the four legs of technology, organisation, environment and the job, but noted a typical lack of integration of approaches to design. She then considered the Maslow hierarchy of needs in the context of design.

Taking a historical perspective, she noted a pendulum tendency in the Finnish approach to design and innovation. The same shell structure could be, and often is, subjected to a series of internal reorganisations.

She gave an account of what had been a promising case study with L.M. Ericsson, comparing cellular and open groups, but consent for the study was withdrawn by the company after a sudden and unpredicted corporate reorganisation.

Paul Vos: a Government Building Agency

Paul Vos of the Dutch Government Building Agency presented an account of the work of a government building agency, evaluating the effects of office innovation. He noted that needs were changing. Information technology changes have brought mobile equipment and electronic communication, and have enabled collaborative working to become independent of time and space. He argued that teleworking remains a marginal activity, but that work is increasingly judged on outputs, and that success of new work organisation depends on trust. With flexibilisation of work processes, communication gains in importance. This raises questions of teamworking within flat organisational structures.

All of these changing needs have impacts on the environment. **Paul Vos** considered issues of place, space and use with respect to the individual, who might be assigned personal or shared space, or left to operate non-territorially.

The final section of the presentation concerned evaluations, in which the effects of office innovation are measured in terms of effects on productivity, quality, flexibility and innovation. What models should we be using?

Lisbeth Birgersson: Facilities Management for Small Businesses

Lisbeth Birgersson of Chalmers University gave an account of a case study of developing space for small businesses in an old former mill, within the growing field of reusing old industrial buildings and offering supported workspace within a managed environment. Apart from supporting individual firms, the work could sustain local communities whose economies had undergone change. She argued the case for principles of justice in local authority planning, for the importance of nearness and values of caring, and for developing concepts of social capital. In her presentation she also summarised the work of a colleague on entrepreneurial facilities management, talking of a balanced incremental development process.

Jesper Steen: Physical Planning for Growth of Economic Life – the Development of a Spatial Analytical Model

Jesper Steen of the Royal Institute of Technology observed that over his twenty years experience of working life research, demand has changed. In the context of large scale unemployment, it is important to improve working conditions in small and medium sized enterprises, which are seen as the potential source of new jobs. The research task has been to develop an analytical model to describe and help us understand urban areas and plans.

Jesper Steen drew on the work of Hillier in arguing that social patterns are influenced by spatial relationships. He offered schematic patterns which characterise rural, urban, city and suburban areas, and added an activity perspective, determining what might happen within the spaces. He was working on a case study of central Stockholm.

Michael Fenker: a Corporate Marketing Case

Michael Fenker described a case study of space design to meet the needs of a corporate marketing department, illustrating issues of corporate understanding of space. The client department was characterised by considerable interaction of autonomous professionals, who operate in an informal style. This results in an experimental site from which we can learn.

A critical move in the project was to reorganise from eight teams to three categories of personal job profiles: experts, integrators and implementers, each depicted as having different space requirements. These profiles, and the corresponding space designs, were linked to performance criteria and stated company goals.

Shauna Mallory-Hill: the Value of Space

Shauna Mallory-Hill is a Canadian architect, now working at Eindhoven. The focus of her doctoral research is on computer tools, linked to the workplace. She is seeking to clarify both problems and design rules. She argued that many work environments are of poor quality, due in part to lack of technical performance knowledge. Design should not be linear, but should benefit from feedback mechanisms. She introduced the concept of performance, with descriptions of outcomes. Working from sources such as journals, workplace evidence, archived material and engineering evaluation tools, she has been developing a case-based hybrid Knowledge Based System support model. Underpinning this is a value relationship model, based on factors such as visual comfort. She is currently developing a portfolio of cases, documenting client requirements and technical performance, and modelling key performance criteria.

Jesper Steen: Office Case Study

Jesper Steen reported on an office case study, derived from the concerns of real estate owners for the costs of frequent remodelling of office buildings used by government departments. Interviews have been conducted with government authorities, with a view to investigating core business needs. He gave accounts based on form and meaning, and on form and function, in which he considered power, rationality and social factors. He noted that amid pressures for rationalisation and improved space for action, we need to ask for whom the changes are made. Who benefits from rationalisation and standardisation? How are these concerns reflected in spatial structures?

Workshop Session

The four sub-themes addressed by the smaller workshop groups were:

1.   Spatial Aspects of Working and Learning (day 2)

2.   Urban Space and Small Enterprises (day 1)

3.   Design and Work for Change (day 2)

4.   Facilities Management (day 1)

Each had been the focus of papers submitted in advance, read by small group members, and available on the workshop home page.

1. Spatial Aspects of Working and Learning

Jan Ahlin: Cooperation Between Universities and Local Communities

Jan Ahlin, from Chalmers University, saw his topic as crucial, and as derived from, but not exclusive to, Sweden, where the economy is stagnating. Over twenty years income levels have slipped from 4th to 19th in the OECD, and Finland has overtaken Sweden for the first time. Sweden is lowest in Scandinavia. There is growing awareness of a crisis. What are the ways out? This is not an entirely new phenomenon. Demand is weak. Wars are one approach (but out of the question today as a deliberate policy), another is increasing equality (difficult in an open market economy). This situation is new. The only practical approach is to increase the knowledge level in society; thus the current policy of university expansion, and the buzz-word "knowledge society". This is not just a craze, but should be taken seriously. Knowledge is becoming the key factor of production, with universities in a key role. The old universities of Uppsala and Lund were the only ones in Sweden for a long period, and the expansion has been in this century, with an additional set of new universities in the last twenty years, involving regional university colleges.

He identified similar practice in Norway, UK and Finland, etc. He argued for comparative research on the roles of these regional colleges, and has spoken to researchers in the USA and Poland. Local universities should be seen as receiving stations for international research, which requires local research activity. Local universities are surrounded by collaborating institutions, which can benefit from distribution of research results, often translated via the universities. This model has physical implications for designing and building universities.

Goran Lindahl: Workspace and Learning

Learning is a classic buzz-word, as satirised by **Tom Marcus**. The spatial dimension has often been omitted from discussions. Global competition increases the urgency of addressing learning issues. Learning is in general positive, but there are complex issues. Organisations and individuals may have different interests. The real underlying issue may be communication. He discussed learning in working life and the workspace.

Goran Lindahl referred to Maslow's triangle of needs, and approaches to conceptualisation. Learning operates at several levels. The workplace provides the context, but also has to support individual developmental learning, and non-routine work. He considered the object and process aspects of space. Individuals may not be free to move around, or to restructure their physical environment, which could help communication and learning. It is important to be able to affect one's own workplace.

There are different types of learning: reproductive, method-based, problem-based or creative. In reproductive learning tasks, method and result are given. In method-based learning, the result is not known in advance. In problem-based learning, only the task is given. In creative learning all stages are open. He discussed how these types of learning may be reflected in the workplace.

He raised two questions. He argued that basic levels in the physical environment are necessary for learning: the workspace should be given over to employees. He then argued that chance encounters are important, and should be supported by design. The second question may be seen as linked to the Third Task context set in the previous presentation.

Colin Clipson: Exploring Worklife and Workplace Issues

Two separate systems are identified, the social and the technical, which have to be brought together. Human systems are older and stronger than technology. Colin Clipson has explored perspectives on designing for change. We need to understand how people really work, not the official story. He quoted John Seely Brown, guru of intelligent systems at Xerox, and Michael Hammer, guru of business process re-engineering, both of whose views have undergone belated change in recent years. He divides work between the bureaucratic and the vital model, implying that the latter is better. Changes in modes of work are from private to public, fixed to mobile, static to dynamic, certain to uncertain. We need to understand the changes. Colin Clipson ended by putting workplace requirements in a Maslow triangle, emphasising the need for groups to be able to control their own environment. This raises questions about the use of individual competence in the context of uncertainty.

A second paper, by Colin Clipson with Kate West, addressed research issues and tools for analysis. What, Tom Marcus asked, are we learning? We are learning about methods of observing the world in which we are interested. Our tools are limited.

2. Urban Space and Small Enterprises

The group's discussions were based on the Chalmers University critique of architectural work, and included reflections on development projects in which members had been involved. Discussions about spatial arrangements are often a useful means of discussing other topics. In SMEs, architects are frequently ignored and excluded, yet spatial relations may be extremely important for business success. The group considered possible links between architectural approaches and modes of learning in organisations: this would be considered further in a later workshop session.

3. Design and Work for Change

Tom Marcus: Machines and People

Tom Marcus was puzzled about the original title of the workshop, and the involvement of "production and work". Either, it is suggested, we have production without work, or work without productivity, and neither of these is sensible. Work is the key focus point, together with workspace, rather than production and production space.

The words have different derivation: production is more abstract, and reflects a systems view. Work reflects a life view, linked to work ethics, work shyness etc. We consider different agents: the individual or the enterprise. Volvo is a producer, not a worker. Production is global, while work is local, dominated by space and time in the particular location. This raises important issues. The interconnectedness between global and local is important.

Much of our rhetoric is about human properties: education, learning etc. Machines cannot acquire these things, yet in our decision processes and thinking about design it is systems aspects which come to the fore. **Tom Marcus** argued that the real role of Sweden in 2001 is to re-emphasise the work aspects rather than production, in a world that has become unsympathetic to the Swedish approach to people, the environment and ecology. Europe, the Far East and the USA have forgotten. Sweden must help us remember.

What is important about space? There are issues of common interest, including movement freely by chance through space. This cannot be programmed. Random chance encounters in space encourage learning. Space helps create solidarity, and the learning and knowledge creation dimension. Internet exchanges have not been a substitute for a physical meeting in a space in Brussels. Information and communications systems increase the demand for such spaces. Space needs will not disappear, but change. People suffer from working in isolation, from spatial contact deprivation. The new technology is not an objective given, but a social phenomenon. **Tom Marcus** argued that we must remove the built environment from the technical sphere of production to the linguistic, social and psychological sphere of work.

Rob Teunissen: Images, Interests, Implementations
A Successful Approach Towards Creating High Performance Workplaces with Administrative Organisations

Rob Teunissen announced that, as a consultant with the Dutch Government Building Agency, he would use the full set of buzz-words. He started with added value, linking to systems which are in constant change. From a business perspective, real estate is simply a facility in space. We need to leave our bodies somewhere.

He considered Duffy's simple model of a grid in which to locate organisations and derive potential design solutions: this is valuable for consultants. Involvement of staff is important, measured in 14 recent projects. However good you are at changing contents, you must be good at the process: you need to be good at both. Frank Becker has argued that facilities are not enough: corporate finance and human resources are also involved. This can mean joint non-core management. Costs of churning can exceed costs of training. Corporate infrastructure debates follow similar lines. These systems are always changing. Success depends on management describing effective behaviour after proposed change. This works.

In the real estate business, people work in teams for finite periods, without support accommodation. New facilities were developed, linked to a research project. The hypothesis was that processes, teams and support were understood, but new ingredients were made available, at cost. The hypothesis is that the work processes would then change, and they do. This is both research and profitable. Pilot environments include furniture on wheels, changeable by staff at cost. Adding money gives something to measure. People do not like spending on space and furniture, but they spend on IT, catering and express mail. How can this be financed? There are at least two important sources. There is unused capacity in current workplaces: management workstations are sitting idle for 80% of the time. 36,000 government officials in The Hague all travel at once daily, adding to congestion.

The challenge is adapting to new strategies: a new approach is arising, starting with management awareness. The problem must be expressed in business terms: goals, process and commitment. Next the level of ambition is set: reducing inefficiencies, providing better support for work processes, creating conditions for change, and creating a learning organisation, using the experience as a case study. Clear communication is vital, followed by elaboration and concept building, developing behaviour. Powerful images are used in work situations. Managers form concepts in their minds, transferring responsibility to the groups concerned. Next we move to the level of facilities needed, and how people can be reached and supported (e.g. help desks, catering). The architect only comes in at the fifth stage, once the owners have developed the concepts. The architect translates this into space. We need creativity, with atmosphere and image: too much money can cause problems! The sixth phase is evaluation, which has to avoid invasive observation.

Key principles are enhanced communication, working in teams, increased autonomy, variable organisational borders, productivity, flexibility and corporate culture.

Rob Teunissen argued for close management to restrict the ego of the architect.

Saddek Rehal: Communication of Insights in Early Stages of Collective Design Processes

Saddek Rehal is concerned with methods and tools, primarily pictures. The actors are required to take their own pictures. The collective design process is a learning process for the participating actors who design themselves by designing their environment. He sees the design process as:

actor – process – artefact

The key issue is the communication of ideas. Within a company there are different disciplines, different language games. Architects communicate with pictures, while others use words. Individuals choose pictures, in a way which has been closely observed in his research. Representation of concepts is complex. Problems can arise at a number of levels:

- articulating the individual insight, using images
- communication within a specific language game
- communication between language games
- communication between language games specific to the company and the architect

The architect can block the imaginative capacity of other agents, so should be kept out of the way.

Turid Horgen: Learning from Success

Turid Horgen asked about learning from success. She showed illustrations of an open conference room in an office environment, in Webster, upper New York State, an experiment in redesign. This became a case study for Xerox labs. The labs produced new patents and accelerated responses. Was this simply hype? The case was discussed at an MIT conference.

How do you learn? The Xerox project was a Laboratory for Remote Cooperation. First it was necessary to understand work face-to-face. A number of situations were considered.

A changed environment means changed work processes; a departing research worker, wanting to take the method with him; a new manager; and a departing manager, wanting to take the method. Eventually the laboratory becomes part of the experiment itself.

A series of seminars were organised, on barriers to change and catalysts to change. Stories were built around each situation. Clashes, territoriality, and proximity themes are raised. External interest has extended consultancy to become teaching.

4. Facilities Management

There was spirited discussion about the nature of facilities management, which had been one of the core themes in the call for papers that had been sent to European schools of architecture.

From the point of view of architects, facilities management is the business of managing buildings. Webster's dictionary offers a wider definition in terms of facilitating the working of the organisation. From the point of view of general managers in the United Kingdom, facilities management can be a means of reducing costs in non-core areas, often transferring responsibility to others, and implicitly downgrading the value of such activities, which are no longer seen as professional but routine. In the UK this has often been achieved through outsourcing, or as part of the process of privatisation. After a period of months, it frequently becomes necessary to relearn how the functions are to be performed, whether by the contractor or the staff of the organisation itself. There were accounts of more principled approaches in Finland and the Netherlands.

It was argued that, like Total Quality Management, Facilities Management represented a fashionable concept that had travelled far and fast, without the benefit of underpinning research.

In those cases where architects have been assigned legal contractual responsibility for buildings throughout their life, the business of facilities management has priority, and indeed the architect is seen as having an enduring involvement. In other cases, where professional facilities management has been introduced, an architect may form part of the team. It is essential for architects to understand the language and concerns of business.

Work in the Netherlands is evaluating the effects of organisational changes, including facilities management, but it is hard to arrive at the right instruments.

The debate moved to issues of the improved education of architects, and the need for architects to assist in educating their clients. Buildings are not just functional, but involve image and psychological factors. Managers tend to understand the issues better when buildings are compared with cars, which need to preserve a resale value, can be used for effect and not just for transport, and there are similar considerations of heating, ventilation and IT.

There was then discussion of issues of time and space, and the ways in which different styles of building enable workers to cooperate. The new forms of office, where colleagues may share neither time nor space, appear to place new demands on company culture. Alternatively, such circumstances call for clear and effective architectural statements.

The final discussion was of adaptability and flexibility, as approaches to the design and construction of new buildings. This was stimulated by examples of new office design, and accounts of experience of designing buildings for use by disabled people. Should the building be built as a shell allowing maximum tailoring of internal design, or should it be built with subsequent adaptations in mind?

Overall there was scepticism about prophets selling simple solutions, whether to problems of business strategy or architectural design. Organisations need a shared vision, which is easier prescribed than delivered.

Spatial Aspects of Work and Learning

Joen Sachs of Chalmers University reported on the work of the group. He noted that the interests of the group were in line with some current European research directions. Lifelong Learning, seen currently as Objective 4 for regional education and development funding in the European Union, can be linked to papers from Jan Ahlin and Goran Lindahl, concerning regional universities and learning in the workplace. In addition, Swedish initiatives such as the "Third Task" have echoes across Europe, and in the UK government has proposed a new "University for Industry", based on learning in the workplace.

Design and Work for Change

Idea generation was followed by a need for design control, then the role of the building designer. There was debate on the role of the architect in the design process. There is also a facilitating role of an outsider. The insiders may work closely together, or in isolation. This raises different definitions of the workplace, and the nature of tasks conducted in space. Users are not interested in elevators (Rob Teunissen), but they ought to be (Tom Marcus). Interventions are necessary, but they can be dangerous. The insertion of images can frame the discussion in a particular way. Using images as a means of intervening can mean too much concentration on the visual, at the expense of narratives.

Reflective Presentations

Peter Ullmark

Peter Ullmark is professor at Chalmers University. We cannot wait, and think out what is best. "Continuous cooperative design processes" summarises the theme of the workshop. Change in companies is real. There are possibilities for making changes in a more human way. This means a need for stability somewhere. Architects look for stability in the physical space, but change impacts the physical environment. He argued that the working group needs to be the stable factor, considering social areas in the workplace. It is hard to discuss relations between the professionals, and then with the users.

The first model proposed went from function to structure, via design. The designer needs to find a link between himself, the field, the language games and the artefact. An expanded model added ethical dimensions, and the objective of obtaining qualities in obvious form, with congenial solutions open to discussion. He offered a three-dimensional model, opened out to include data from experience. The three corners have values, data and experience, and the links between them constitute different forms of knowledge: constructive, critical and organised. He then added technical, scientific and artistic knowledge, opening discussion of empirical and tacit knowledge. When discussing cooperative design work, these issues of knowledge arise.

Joen Sachs

Joen Sachs of Chalmers University declared that he had been through this kind of workshop experience before. The results always take time to penetrate the outside world. We have to address external organisations. If this group is seen as comprising researchers, we need to communicate with a number of other groups, and gain acceptance: research society and research groups such as work research, the professionals, users, clients, authorities and financing institutions, and architecture schools. In order to break through and arrive at more valid and usable knowledge, we also have to work seriously with the basics. Given the richness of sessions, the diversity of inputs, and the variety of research perspectives, there is a great need for a research agenda which gives us a common ground for future communication. We have to join forces, and arrive at a common basis for our studies with precise language and clear definitions, a necessary framework for all research of high quality. He reissued a paper from an earlier seminar on Corporate space and architecture. It advocates the creation of international and interdisciplinary glossaries and tentatively suggests two common areas for research: the changing nature of work and the design process.

Conclusions

Thomas Marcus referred back to the original four themes, and noted that there had been little discussion on research methods and techniques. Where we know the authors, we know their backgrounds, but there was little in print on this occasion to provide the foundations for a European research agenda. Is this because we are immersed in the subject, or is it rather a sign of weakness? There is considerable achievement behind the simple statements, but it is hidden. The work needs to be massaged in the coming months, and analysed in terms of research priorities in a way that external potential partners can appreciate. He did not see a research agenda as having emerged.

Joen Sachs argued that the idea had been to present areas of research for discussion, and propose new areas of research. He highlighted the papers by Saddek Rehal, Jan

Ahlin and Jesper Steen, and saw the workshop as raising possibilities, a first step. There have been changes over the past six years: the nucleus has grown, and there is an international research community.

Jesper Steen noted that research has been close to practice, concerned with producing effective results. There may be a need for a more principled theoretical approach.

Tom Marcus argued that not much work on research methods was visible, compared with what authors could have provided. It would be useful to make research methods explicit, and to set out research agendas for new projects. This would provide a corpus of disciplined discourse, from which a discipline could be identified. In itself it would not make good reading. There are no short cuts.

Workshop Participants

Jan Ahlin, Chalmers University, Sweden
Lizbeth Birgersson, Chalmers University, Sweden
Colin Clipson, University of Michigan, Ann Arbor, USA
Richard Ennals, Kingston University, UK
Michael Fenker, Ecole d'Architecture de Paris la Villette, France
Turid Horgen, MIT, USA and Design Matters, Norway
Ritva Lappalmainen, Engle Construction Services, Helsinki, Finland
Goran Lindahl, Chalmers University, Sweden
Shauna Mallory-Hill, Eindhoven University of Technology, Netherlands
Tom Marcus, Strathclyde University, Scotland
Sadek Rehal, Chalmers University, Sweden
Joen Sachs, Chalmers University, Sweden
Lena Skiöld, NIWL, Sweden
Degenhard Sommer, Technical University of Vienna, Austria
Jesper Steen, Royal Institute of Technology
Rob Teunissen, Dutch Government Building Agency, Netherlands
Peter Ullmark, Chalmers University, Sweden
Paul Vos, Dutch University of Technology, Netherlands

Reflections on the Workshop

This workshop involved a distinct community of professionals, viewing the workplace from a spatial perspective. Issues of the spatial environment also arise elsewhere in Work Life 2000, such as in the context of the small business theme, and links could be developed. Some professionals present regarded issues of space as fundamental, underlying all other concerns, while others argued that this approach to interdisciplinary issues tended to lead to architects being marginalised or excluded altogether. There is a need for the different professions to gain experience of working together and developing common language.

This was the first of the workshops to be preceded by an exchange of papers via the Internet, and the development of a workshop-specific Web site. This use of the technology appears to have raised the level of discussion at the workshop, and extended the range of publishing and reporting opportunities.

2. How Does Medical Surveillance Contribute to the Objectives of Directive 89/391?

The workshop leader was Anders Englund, of the National Board for Occupational Safety and Health, Sweden. The rapporteurs were Leif Aringer, NBOSH, Sweden and Karel van Damme, Univ de Louvain, Belgium. The workshop was held in Brussels, 7–9 September 1998 at the Office of the Swedish Trade Unions.

Abstract

The European Union Directive 89/391, among other stipulations, requires that the employer makes sure that employees are provided with opportunities for medical screening of possible disease related to his or her working conditions. The provisions shall conform with national legislation and/or practice, and the opportunities are to be available to everyone who desires to use them. It is further said that the services might be provided by any part of the medical system, and not only by multidisciplinary services.

The purpose of the workshop was to review what medical screening might be beneficial. There are different purposes for early detection programmes in the occupational health context. The purpose is not only to improve survival and quality of life for the worker, but also to serve as a signal system for identification of hazardous working conditions. Although the first principle is to detect and eliminate such conditions beforehand, a second line of defence or prevention is to ascertain that the measures taken have been sufficient and no harm caused.

The workshop invited some distinguished scientists from the EU and the US to address the topic from both a strategic and a clinical point of view. From the clinical perspective reviews were given of the state of the art for early diagnosis of relevant diseases like those of the respiratory, metabolic, neurological and blood forming systems. Certain regulations, such as for carcinogens, demand that medical surveillance continues after retirement age. The feasibility of such programmes, and where responsibility lies, requires thorough consideration. The relevance of pre-employment screening as a tool for prevention of musculo-skeletal diseases is another area of concern in view of recent national legislation against discrimination against the disabled.

Historically there have been different views on the role of medical screening in occupational safety and health in the different member states of the EU. The workshop considered these differences, with a view to reaching a common opinion on ways forward.

Background

The workshop considered EU directive 89/391 Article 14 in detail. It requires that measures are introduced to ensure that workers receive health surveillance appropriate to the health and safety risks they incur at work. The measures are to be introduced in accordance with national law, and the health surveillance may be provided as part of a national health system. It further requires that those measures should ensure that each worker, if he so wishes, may receive health surveillance at regular intervals.

A number of questions arise with regard to definition:

- What is health surveillance? Is it equivalent to medical surveillance?

- What elements of a surveillance plan are "appropriate to the health and safety risks incurred at work"?

- What does "at regular intervals" mean in relation to the above "appropriate to risks incurred"?

Other questions arise from the nature of such "health and safety risks":

- What types of endpoints (diseases) are detectable at a stage early enough for cure to be possible?

- Should emphasis be, for example, on early indicators of disease or biological markers of exposure?

- What are the ethical aspects of such surveillance programmes?

These questions need to be addressed and developed if there is to be a cohesive approach in the Member States of the European Union. The need for such an approach led to the inclusion of this topic among the tasks for the DG-V of the Commission Ad Hoc Group on Multidisciplinary Services and Health Surveillance. Accordingly a group of experts on occupational medicine was convened in Brussels to review which surveillance methods are "appropriate" to a range of potential "health and safety risks". Ongoing national health surveillance programmes were presented, and their benefits and limitations discussed. The ethical aspects of health surveillance must be considered, and they constitute a substantial part of the overall review.

Introduction to the Workshop

Anders Englund introduced the themes of the workshop, using the background paper above as the basis. He saw the workshop as playing an important role in developing European Union policy, in association with the Ad Hoc Committee, on which he represents Sweden. He referred to a recent ILO report by **Greg Wagner**, which had come from a tripartite group, assisted by experts, who were constrained not to contradict current practice by member states. There was a case for a separate discussion of invited experts, free from such constraints.

The first day addressed the European directive, and considered occupational health from the perspective of different parts of the body, together with general ethical considerations. The second day was based on country reports and sessions of two discussion groups. The third day considered the reports from the discussion groups, with the objective of reaching general conclusions. It was hoped that the workshop would end with a draft report.

The European Commission Directive 89/391 Article 14

Alexandre Berlin of DG-V introduced the Commission perspective on Directive 89/391 Article 14, which followed earlier directives dealing with particular substances. He identified two groups of Member States, one of which, comprising France, Italy, Spain, Germany and Finland, were happy to proceed beyond general surveillance measures, while the other group were opposed to the medicalisation of occupational health. The directive had been approved on the basis of general concern for prevention. An Ad Hoc Group had been established to consider multidisciplinary services, including health and medical surveillance. A Danish survey had revealed diversity of current national practices in occupational health services.

Discussing the detail of the Directive in terms of the questions posed in the Background Paper, he noted that medical surveillance was seen within the broader field of health surveillance. The intervals between surveillance opportunities would depend on the risk involved. He highlighted ethical issues that are inherent in the field: the intention is to improve working conditions, not to remove workers on spurious grounds. Occupational health is concerned with workplace factors, and not with predispositions. He noted that there are ongoing debates between the European Commission and the Swedish Government concerning interpretation of the Directive, and the requirement for all workers to have access to surveillance.

Different Directives, agreed incrementally over time, have embodied different approaches to regulations and their implementation across Member States, and it is time to give some overall coherence to the role of medical professionals within the wider set of professionals concerned with occupational health.

Discussion

Marja Sorsa emphasised the importance of separating medical and health dimensions, ensuring proper provision for psychosocial factors and ergonomics. **Vito Foa** described the move from the traditional role of protecting workers against risk, and towards the promotion of health. **Marja Sorsa** saw occupational health as changing, together with changes in working life. **Dan Murphy** reported that Ireland recommends cooperation between occupational medical disciplines, including occupational health physicians, and favours health surveillance undertaken by the worker, who has access to medical expertise through the health service.

Screening and Surveillance

Greg Wagner offered an overview of "Screening and Surveillance for Occupational Disease: Principles, Problems, Promises". Working from the recent ILO report, he offered definitions of screening as being concerned with the individual, able to identify but not prevent illnesses. If undertaken early enough, screening could identify problems and benefit health. Surveillance was concerned with groups, and involves the collection, analysis and reporting of information, usually on a periodic basis. The two are linked, in that surveillance uses test data from screening. Tests that are not useful for screening can help in surveillance.

He offered a diagrammatic account of the history of a disease, locating the key role of surveillance in secondary prevention. Primary prevention is a matter of analysing machines in the workplace, and improving the work environment. New developments in genetics might conceptually be located at the primary stage.

Karel van Damme saw this account as based on disease, and affirmed a positive definition of health as a state of well-being. He noted that having a job could be seen as good for the health, and cited general adverse outcomes that could prejudice health, beyond simple disease.

Greg Wagner argued that a disease had to be important in order to warrant intervention to make a difference. It needs to be sufficiently prevalent to be worth testing for. This provoked some discussion. Screening tests need to be credible, acceptable, have clear cut-offs between normal and abnormal outcomes, good predictive value, affordable cost, and standardisability for consistency, accuracy and reproducibility. He acknowledged that there can be problems with overlapping groups which give incorrect and misleading conclusions. He noted the range of tests available for different diseases, deriving from many areas. For example lung diseases could involve chest imagery, sputum and skin tests. There can be variability between readers of test data, which casts doubt on the conclusions, though this is less true of modern scanner technologies. There is an absence of data showing the benefits of physical examinations for the individual, but some appear to be powerful predictors, such as the Exercise Tolerance Test. It was often a question of applying population data to an individual, which means looking at both high and low prevalence populations. Intervention then needs to be available, acceptable and effective. Actions will include modifying the workplace, educating workers, and commitment to follow-up. Acceptability of intervention has to be seen in the legal and social context, taking into account the knowledge, attitudes and beliefs of those concerned. There are issues concerning notification, confidentiality, ownership of the process, and the facilities available, including training, record keeping and quality control.

Surveillance raises new questions, including the identification of trends, definition of the scale of the problem to be addressed, the identification of new hazards, targeting and evaluation. In the cases of asbestosis and silicosis, pneumoconiosis encountered problems of different national definitions. It was difficult to track causes of death, seldom attributable to one cause alone. European summary data linked prevalence of these diseases with the density of coal mines, while in the USA there was an interesting distortion, as sufferers from these diseases often tend to have moved to Florida in retirement. One approach is to group individual data to give population information, but this may be frustrated if examinations are refused. Case based surveillance can give rise to the discovery of other cases. Success depends on participation, a collective approach and effective dissemination of outcomes. In general it is a matter of controlling exposure to hazards. Thus screening and surveillance need to be integrated with exposure monitoring and control. This raises choices between comprehensive and selective approaches, in which we seek to avoid giving false reassurance or false abnormal test results. There are resource implications.

In conclusion, **Greg Wagner** saw both screening and surveillance as useful. Neither prevents illness, but can contribute to prevention. Screening can provide valuable surveillance data, while periodic examinations can be an opportunity for screening. Genetic testing is different, but must increasingly be seen in the context of overall prevention strategies.

Vito Foa argued that physicians now work with other professionals, and that their work includes health promotion and campaigning. It is not simply a matter of health at work, for the individual's outdoor, home and personal circumstances are also relevant, and overlapping in impact. The individual is the integration point for these different dimensions.

It is possible to be facile in the emphasis on the workplace, or to use it as the venue for general health promotion, rather than dealing with workplace specific issues. There was debate about smoking and lung cancer, and the validity of workplace screening: there is a case for screening high risk groups, but the results could be anomalous if attention is on the workplace rather than the level of smoking.

Arne Wennberg gave an overview of the nervous system, noting that the baby is born with the full complement of 100 billion nerve cells. Once destroyed, nerve cells cannot be replaced. He presented nerve cells and synapses as the hardware and software of the human system. Toxic effects can have long-term consequences in terms of reducing effective functions, but he argued that we start with considerable spare capacity. At the lowest level of impact on the peripheral nervous system, effects can be reversed.

He classified substances into neuro-toxic, not neuro-toxic, and harmless, and described six levels of effects that could arise. He presented the WHO list of research methods, with recommendations as to appropriateness. His past work on alcohol at the Karolinska Hospital gave useful data applicable to other organic solvents.

In response to questions, Arne Wennberg focused on the neurological dimension of screening and surveillance. Testing could not be *in vitro*, and there are ethical questions arising from human tests with neuro-toxic hazards. This is a specialist area, less well known than respiratory work. The methods are highly sensitive, but not necessarily specific.

Genetic Testing and Ethics

Marja Sorsa addressed issues of genetic testing and ethics, on which she advises the European Commission. She identified distinctive characteristics of genetic testing:

- It only needs to be done once per person

- Genetic screening for inherited genes gives information concerning relations

- There can be a long period of latency between discovery and outcome

She gave an overview of the Human Genome project, which is due to give a complete mapping of the genes by 2003, and has provided a vast volume of information. There are clear commercial issues with relevance to the work environment. Testing kits are being developed, but of variable quality. Rules and codes of practice are needed.

In the workplace the issue is genetic monitoring, considering effects of exposure, identification of hazards, assessment of risk and management of risk. Screening can take place before or during employment, looking for inborn factors and possibly excluding applicants. A few serious diseases and metabolic deficiencies have been identified as clearly correlated with a single gene, or a couple of genes, but correlations have been slow to emerge: this is perhaps not surprising as there are 80,000 genes, and many problems are due to interactions of several genes. Familiar examples fall outside occupational health (such as breast cancer). Tests are becoming available for inherited diseases, and this is big business in the USA where over a million people are potentially affected by diseases such as sickle cell anaemia, cystic fibrosis, Huntington's chorea and haemophilia.

A number of international organisations, including the European Commission and Council of Europe, have addressed the ethical issues, seeking to restrict genetic testing to health purposes. The UNESCO declaration on the Human Genome is not legally binding, but emphasises human rights, that benefits of genetic knowledge should be made available for all, and that we all share the same gene pool. **Marja Sorsa** is a member of the EU Group on Ethics in Science and New Technologies (EGE), whose work deals with biotechnology, information technology and social values. She set out the basic ethical principles of beneficence and non-maleficence, decisional autonomy and justice. She highlighted differences in risks, and the presence of competing interests. The nature of scientific uncertainty is complex and poorly understood. Knowledge is unevenly distributed, and much action is unconcerned. Genetic tests require informed consent: this requires the information to be right, understood, authentic, and without conflict. We should be adapting workplaces for workers, not adapting workers for workplaces. Genetic tests should not be used to exclude workers who had disease risks that are not job related. Diseases should be prevented by proper job placements. Genetic tests should be used to predict diseases in a medical setting, but not leading to the exclusion of tested workers.

Ethical Issues

Karel van Damme spoke from experience of the European project on genetic testing for workers. In general terms: a person should be declared fit for work; unless the opposite decision would be a medical mistake, in light of evidence of a threat to the health of the individual and others, which was not preventable at the workplace by other means, on the basis of current knowledge and best practice. It could then be the duty of the employer to offer another suitable job. Issues remain as to who decides, and on what grounds.

He presented data on screening for allergies, showing that preventing one potential case had involved numerous exclusions and false diagnoses. He looked at issues of susceptibility and exposure, showing that at high levels of exposure to cigarettes, susceptibility was less significant. Overall, health issues are bedevilled by scientific uncertainty. Who will use and interpret the tests? Each must be seen in context. In general, the stated ethical aims of the European Union are respect for human dignity, social justice, solidarity and democratic participation.

Occupational health involves working conditions, pre-job testing and health surveillance. Genetic testing has raised issues of regulations, organisational frameworks, practices and professionalism. Occupational health professionals are committed to the right to work, to protecting health, and to enabling employment. We face the challenge of standardisation brought about by the free market and globalisation. This may increase pressures for a reductionist approach, with susceptibility testing and the setting of biological limit values. In Europe these matters are handled on a tripartite basis, but we still have to determine who decides what, how, why, and for whom.

National Policies and Practices

Ole Svane, Denmark

Denmark has a free health service, with no costs for hospitals or GPs, and no private medicine. **Ole Svane** set the scene of Denmark as a small agricultural economy, with light industry and services. Health surveillance is regulated, and the set of EU directives has been transformed into Danish regulations. Citizens are always free to go to the doctor. Employers are denied access to health data on their employees, but can obtain advice through government agencies. There is a small team of 30 full-time occupational physicians.

Frank van Dijk, Netherlands

The presentation focused on curbing pre-employment assessments, which developed in the Netherlands to meet employer needs, which include reducing the cost of sickness absence. 1000 physicians had previously been linked with insurance companies, and became a commercialised Occupational Health and Safety system. Health selection has been a key issue, despite legislation and codes of practice. Many companies are not insured.

Dan Murphy, Ireland

The Irish legal system is similar to that of the UK, with a similar, but less complete, national health service. Particular problems are presented by foreign firms, with implications for surveillance, and by small firms. Most occupational physicians are part-time, and national policy favours multidisciplinary teams, offering skills at a number of levels in the field of occupational health. EU Directives have been implemented as regulations, but a further complication comes from civil law, under which workers can sue for compensation.

Catherine Bonnin, France

The French system, little changed since 1946, is driven by principles of being obligatory, preventive, and the responsibility of the employer. Firms either maintain their own medical services, or form inter-enterprise partnerships. There are 6,000 occupational physicians, with some 200 working in the construction industry. Special attention is given to particular groups of the population, and to particular professional hazards. Catherine Bonnin outlined the system in France, covering the Ministry of Labour, the employer (who is obliged to consult the physician) and social security. Ethical problems arise from mixing prevention and compensation, and the responsibility of the employer may place the physician in an invidious position with respect to the worker. There are annual examinations, but an absence of overall analysis of the wealth of data. Thus, although physicians observed signs of the hazard of asbestos, companies were able to resist pressure to act.

Dolores Sole, Spain

OSH legislation in Spain is based on two regulations. In some cases health surveillance equates to medical examinations, but when we examine the specific functions of medical staff in multidisciplinary OSH preventive services, we find a broader view. There are two dimensions in health surveillance: health surveillance targeting workers in companies, and health surveillance as a part of national OSH policy. Legislation covers the provisions of EU directives, and a national information system is being developed by public health administration specialists. Organising the resources to develop preventive actions could be through a mixture of four approaches. Two refer to occupational preventive services. Company teams need access to at least two specialisms within occupational preventive disciplines (possibly held by the same professional if certified). External teams need four specialists.

Vito Foa, Italy

The Italian system is based on over a century of legislation, and a national system of about 6000 occupational physicians, many of whom are not full-time. Practice varies across the different regions of Italy, in terms of practitioner activity and insurance provision, but in principle offers a system based on medical surveillance.

Gerhard Triebig, Germany

The German system is based on a combination of health and social insurance, undertaken by accident insurance carriers. There is a large scale process of preventive examinations, the majority of which show no health risks. There are problems with Small and Medium Sized Enterprises. The key emphasis is on risk identification. Examinations cannot be compulsory, due to legislation on self-determination. Key priorities include education of physicians.

Discussion Groups

The workshop divided into two groups, discussing

- role of health surveillance in workplace safety and health – limitations and opportunities
- national policies and practices – weaknesses and strengths

In each case ethical considerations were to be considered in an integrated manner.

Health Surveillance: the Article 14 Debate

The group defined surveillance in terms of ILO definitions and the wider picture. They then sought to identify appropriate features. The use of words was vital, using choice rather than selection, and maintaining the ethical position set out by **Marja Sorsa**. National practices vary, and the wording should encompass the range of good practice. For example, autonomy should be placed above paternalism, preserving the voluntary principle as, for example, specified in German law. At the present time certification is required by law. Appropriateness of intervals and regularity depends on the risks incurred, for explicable practical reasons.

The debate on Article 14 focused around modalities and regularities, concerning the interactions between national and European level activities. Are all workers covered? What are the details of health surveillance? The full text of Directive 89/391, Article 14 is as follows:

1. To ensure that workers receive health surveillance appropriate to the health and safety risks they incur at work, measures shall be introduced in accordance with national law and/or practices.

2. The measures referred to in paragraph 1 shall be such that each worker, if he so wishes, may receive health surveillance at regular intervals.

3. Health surveillance may be provided as part of a national health system.

The text reflects the compromise struck in 1989. The European project has progressed since that date, and agreement may now be possible on a supporting set of definitions and interpretations, supported by a rich range of documented case studies. Specialist accounts could be given of elements of health surveillance that are critical to appropriateness. This encompasses ethical and practitioner issues, and must cater for the needs of the citizen as consumer.

Article 14 needs to be read together with others in the Framework Directive, such as Article 7, concerning Protective and Preventive Services, and Article 11, concerning Consultation and Participation of Workers. It was assumed that in Article 14, Paragraph 1 addresses employers, who have the duty to make a risk assessment. Paragraph 2 is seen as addressing the worker, who is entitled to receive consequences of risk assessment if he or she chooses. Defining health surveillance as continual would render Paragraph 2, stating the entitlement to receive it at regular intervals, a tautology, apart from the stressing of voluntary involvement. Paragraph 3 relates to the means by which government should discharge responsibilities. Mechanisms for decisions are left to subsidiarity.

It would not be normal English usage to refer to workers as "receiving health surveillance": the verb "receive" is applicable to "medical screening" or "individual examination". It may be that such ambiguous interpretations of the process of health surveillance underpin Paragraphs 1 and 2. It is hard to define what would constitute voluntary participation in practice.

There could now be different views on subsidiarity, as it applies to this Article, in light of the enhanced Social Chapter, to which all European member states have now subscribed, and the employment guidelines, now embodied in action plans by all member states. In order to move the debate onwards, it may be necessary to build some scaffolding, linked to Article 14, in terms of the state of the art of Occupational Health Surveillance.

This means that the report from the workshop to DG-V has to work on different levels, involving a combination of textual annotation, professional guidance, and case study examples.

National Policies

Health surveillance was seen as part of programmes of preventive action, with a focus both on individuals and groups of workers, with the purpose of characterising the health risks for both individuals and groups. The consequences may be seen on a continuum between exclusion from the workplace and a complete adaptation of the workplace to workers, but a clear account can be given of the objective, which is to maintain or improve the health profile of the group. This is achieved through identification of possible risk and the population concerned; planning action; implementation; and evaluating the outcomes, considering both hygiene and health together.

When judging the appropriateness of a health surveillance practice, it is of primary importance that occupational risks are being addressed.

Directive 89/391 deals separately with key issues:

- Article 7: Protective and Preventive Services
- Article 11: Consultation and Participation of Workers
- Article 14: Health Surveillance

The Directive does not specify that Article 14 is subject to the involvement of workers' representatives, because Health Surveillance is described as an individual right, and not part of preventive action. The group viewed this situation as unfortunate. Health surveillance should be subject to consultation and participation of workers' representatives.

The group then considered *minimalistic* and *maximalistic* options for occupational health services, abstracting from the country presentations given earlier.

Under the *minimalistic* option (for example Denmark, Ireland and the United Kingdom) highly exposed groups are identified, and the end points which will be addressed through the health surveillance programme are also clearly identified. A rational testing procedure is elaborated, together with the information which will be given to the individual worker and the possible changes at the workplace which may result. The efficiency of the changes in the workplace may be evaluated with an appropriate periodicity using the same testing approach. This approach concentrates on the main issues by asking questions about risk, undertaking risk assessment and exerting effective control. This allows quality assurance in whatever health surveillance is carried out. The planned approach allows the effective use of team members with different skill levels to undertake particular actions as directed. It is efficient in terms of the use of labour and financial resources.

Under the *maximalistic* option (for example France, Belgium, Italy, Spain and Germany) all workers are entitled to have routine health surveillance in the form of a medical examination by a competent occupational health physician who is familiar with the workplace, possibly on an annual basis. This is done with a view to discovering both collective and individual health problems, which are not necessarily limited to a particular occupational risk. The aim is to use the medical examination to contribute to improving both collective and individual health protection, and to protect the right to work of every individual, explicitly including vulnerable workers. Rather than "surveillance", the medical examination takes the form of an exchange between the occupational health professional and the individual worker. This is presented as a total health approach in countries which do not have free national health systems.

Given these two contrasting options, with long histories and traditions, a number of general points were agreed:

- It is critical to link the surveillance system with the environment.
- Lack of a real surveillance system may lead to the detection of adverse health effects in individuals when the degree of incapacity has reached a more advanced stage. This presents particular problems for people with increased susceptibility.

- The solution appropriate to deal with a particular occupational health risk may differ in time and place, since technologies, settings and populations are different, or change over time.

- Practical solutions should include the planning of several alternative strategies in order to optimise the effectiveness of the contribution of health surveillance as part of a preventive programme.

Thus we will maximise the identification of early health effects, and the planning of preventive actions, and have a basis for evaluation. It was agreed that health surveillance should be undertaken by properly skilled people specialising in occupational health, and with a specialist training in risk communication. Traditions vary in the different countries, with particular implications for the roles of occupational health physicians with respect to other skilled professionals.

Workshop Participants

Leif Aringer, NBOSH, Sweden
Alexandre Berlin, DG-V, Belgium
Catherine Bonnin, APMT, Paris, France
Karel van Damme, Univ de Louvain, Belgium
Frank van Dijk, Coronel Institute, Netherlands
Anders Englund, NBOSH, Sweden
Richard Ennals, Kingston University, UK
Vito Foa, Univ of Milan, Italy
Dolores Sole Gomez, CNCT, Spain
Dan Murphy, HAS, Ireland
Lena Skiöld, NIWL, Sweden
Marja Sorsa, Ministry of Education, Finland
Ole Svane, Dir. Arbeidstilsynet, Denmark
Gerhard Triebig, Univ of Heidelberg, Germany
Gregory Wagner, NIOSH, USA
Arne Wennberg, NIWL, Sweden

Reflections on the Workshop

The content and conduct of the workshop was fully in line with the original objectives of Work Life 2000: it involved a small international group of experts, building on many pre-existing connections, and was able to develop a programme organically rather than depending on lengthy formal expositions. It was genuinely a workshop, in that the European Commission requested advice and support on a difficult set of issues, and time was devoted to small group discussion on the wording of European Directives, in light of definitions from ILO and WHO. The support facilities of the office of the Swedish Trade Unions meant that key documents could be copied.

The workshop group were predominantly senior occupational health physicians. They were dealing with a subject that involves professions allied to medicine, in addition to their own specialism within the medical profession. General consensus was achieved on the governing medical principles, and on the ethical framework within which health surveillance should be undertaken. Issues remain concerning medical professionalism, and the perspective that this gives to discussions of skill, qualifications and confidentiality. Some of these issues will be considered by the Ad Hoc Group on Multidisciplinarity and Health Surveillance, where it may be harder for members to transcend their own national settings.

The wider context of political and social policy cannot be ignored, not least because the countries which were the key examples of minimalistic occupational health strategies are also proud of their free national health services, which provide an underlying context of entitlements to medical services. In other words, there are many routes to "total health", and, in good European style, "no one best way".

The workshop process can be seen as reflecting the nature of Europe as a Development Coalition, as movement continues towards shared consensus policy among expert professionals, despite the diverse backgrounds and histories of member states. In this context, I see the deliberations of the workshop, and its mode of conduct, as falling within the field of Work Organisation as set out by Bjorn Gustavsen, and as implemented in policy by Allan Larsson as Director-General of DG-V at the European Commission. It involved discussion of the different experience of member states, responding to an invitation from the European Commission to facilitate policy development and implementation. There were contrasts and complementarities between national situations, into which a single set of European Directives has to be implemented. Work Life 2000 is a Swedish initiative, but the country case presentations were from other member states, with a valuable expert input from the United States.

The workshop was linked to a practical agenda, concerning Directive 89/391 with the European Commission, and participants were authoritative representatives of member states. While being primarily occupational health professionals, their depth and breadth of experience gave them insights into the practicalities of policy development and implementation. The discussion was specialist in focus, but should be made accessible to more general readers of the Work Life 2000 yearbooks. For example, the scientific insights into genetic testing give added impact to the ethical conclusions. The detailed debate on a particular European Directive gives insights into relationships between member states and the European Union.

There are very practical conclusions for DG-V from this workshop. In the emerging situation in the European Union we need to find a way of moving from our different experiences in national health services, which cater for individual personal health and occupational health. Multinational employers have employees in numerous countries, with different occupational health regimes, but involvement in the same industrial processes. This is now a central issue for trade unions, employers and governments, part of the "level playing field". It has implications for the import and export of jobs, with associated occupational health features. In essence, we seem to require a framework for a virtual European Occupational Health Service, interfacing with national health systems. The framework can be filled out with networking across Member States. The alternative is to maintain the unequal balance of knowledge and power between the social partners, with inconsistent insurance support of last resort from the individual Member State.

3. Environmental Management and Health and Safety

The workshop was organised by Lars Grönkvist, Henrik Litske and Gerard Zwetsloot, and led by Henrik Litske. It was held on 3–4 December 1998 at the European Foundation for the Improvement of Working Conditions in Dublin, and organised in association with the Swedish Government, the Swedish National Institute for Working Life, Arbetar Skyddsnamnden, Arbetarskydsstyrelsen and NUTEK.

Introduction

Dr Hans van Weenen was chairman of the workshop, which he set in a context of continuously improving events concerning sustainable development. He welcomed the diverse backgrounds that were represented. **Ingvar Söderström,** Director of the National Industrial Safety Board, provided an introduction to the Work Life 2000 conference in January 2001 and the preparatory workshops. The challenge would be to implement the new knowledge that resulted from the extensive international process.

Birgit Erngren, General Director of NUTEK, introduced the work of NUTEK, one of the sponsors of the workshop. NUTEK takes a market-based approach to technology and its applications, and places great emphasis on analysis of the effects of activities which it is asked to support. One role is as Swedish coordinator of the Environmental Management and Audit Scheme. NUTEK can point to step by step improvements as a result of its work; its holistic approach is concerned with sustainable development. She described work on eco-design, and collaboration between major automobile manufacturers in developing the Green Car. Preparations are well under way for a major conference in Stockholm in June 2000, with the theme of Eco-Efficiency.

Eric Verbogh, Deputy Director of the European Foundation, was glad to see preparations being made for the new Millennium, drawing on a breadth of expertise.

Henrik Litske introduced the research strategy of the Foundation concerning sustainable development. The Foundation involves Employers, Trades Unions, Governments and the European Commission, and the set of challenges for the current four year period includes sustainable development. This could be justified in terms of the needs of the planet, commitments set out in European treaties, and the importance of sustainability as a theme, as set out in the Brundtland Report and in the Fifth Framework documents of the European Commission. Sustainability has to be seen from environmental, economic and social perspectives, and in terms of working conditions. It is important to be grounded on examples at company level. The focus of the Foundation is on the needs of the social partners, on developing instruments rather than on monitoring, with a medium or long term and integrated approach. High priority projects include:

- Design for sustainable development

- Professional education and training
- Economic and fiscal instruments

This involves the following initiatives:

- Identification of sustainable development networks
- Practical examples of sustainable design in SMEs
- EU's environmental management and audit scheme
- Renewable resources
- A tool box for the social partners

Over the coming two years the emphasis would be on:

- Support systems for SMEs and micro firms on sustainable development
- Health aspects of workers dealing with sustainable production and services

This would include further improvements in the Web site (http://www. eurofound.ie/), and a dissemination programme which includes the present workshop, and has the objective of assisting practitioners at a local level.

Gerard Zwetsloot
Best Practices and Challenges in the Perspective of Sustainable Development

Gerard Zwetsloot referred to his recent publication with the European Foundation "Design for Sustainable Development: Environmental Management and Safety and Health", and provided an overview of the field at policy level, company level and at conceptual level. Sustainable development is a popular slogan, but rarely followed through in policy development or implementation. He summarised the situation regarding EMAS and ISO 14001, where requirements are specified and certification is available, with OHSM where there are guidelines but no specification and no certification. ISO 14000 now has 5700 companies registered worldwide, with the lead taken by Japan, Germany, the UK, Taiwan, Sweden and Netherlands. Environmental Health Management Systems deal with discrete unplanned events (seen as safety issues) and with normal continuous events (environment and health). There are links with work on Quality Management Systems, through the quality principle of striving for excellence, and the aspiration to do it right, even for the first time.

The Rio Declaration in 1992 had placed human beings at the centre of concern, and stated an entitlement to productive life in harmony with nature. The Brundtland Report had stress the importance of the human use of scarce resources, and set a context of safety, health and the environment. From this tradition has come work on stakeholding, and a recognition of the importance of external relations and communications: this encompasses customers, suppliers, neighbours, BGOs, workers, trade unions, governments and, of course shareholders.

At the policy level it is important to:

- Stimulate and reward companies developing EMS
- Involve social partners in OHS

● Coordinate different aspects of sustainable development

At the company level, sustainable development needs to form part of corporate strategy, infusing the core values, mission, core technologies, other core competencies and the process of business planning (including mergers and take-overs). It needs to inform the evaluation of business opportunities, Human Resource Management policies, strategic marketing and corporate image. It is apparent that simpler management systems are required for SMEs, integrated with product stewardship. This will be the subject of a further workshop.

There are important organisational concepts, such as continuous improvement, adaptation to changing circumstances and the development of learning organisations. Technology concepts include green chemistry, process intensification and sustainable product development. As for people, there are issues of leadership, commitment, provision of a supportive atmosphere, involvement, participation, motivation, creativity, learning (both individual and from one another), cooperation, communication and company culture. Japanese experience has shown the benefits that participation brings to quality. Overall there are a number of generic concepts: sustainable development, product stewardship, organisational learning, stakeholder management and social responsibility. Rather than emphasising terms such as environment, safety and ISO 9000, it would be more useful to stress healthy work, sustainability and excellence.

Hans van Weenen noted that in July 1998 only 1738 companies in Europe were registered for EMAS: some 0.9%. Should we regard our worthy discussions and sets of concepts as a buffer against change or as an incentive for change? He recalled that 90% of companies in Europe are SMEs, where management systems might be at the human level of a discussion over breakfast.

Erik Asplund
Environmental Management Systems in Small Companies

Erik Asplund argued that EMS are potentially effective tools for SMEs, but, like vacuum cleaners, they need to be used. This is one of the concerns of NUTEK, who have been seeking to bridge the information gap, dealing with 2000 SMEs per year, in areas relating to EMAS and ISO 14001. Seminars provide a powerful means of involving SMEs, who are then supported in networks, of which there are now 40 with 500 SMEs. Grants are available, and specialist support at national, regional and local level. SMEs need easy tools to get started, network links with other companies, and seminars as a proper forum. Above all, the cost of involvement needs to be affordable.

The approach with SMEs has shortcomings, and it is hard to quantify the benefits. Many companies can be scared by the apparent short-term costs, and find it hard to move from reactive to strategic modes of management. However, networking works, and brings benefits to those who continue.

Hans van Weenen praised this positive approach, but noted that the number of companies involved in EMAS was still no greater than the number of delegates attending conferences on the subject. The challenge is dissemination. He recommended a number of recently published books.

Jan de Saedeleer
An Environmental Management Framework at Procter and Gamble

Procter and Gamble operates in 140 countries, with 120,000 employees, and over 160 years has developed a complex internally organised system. Safety is critical for the survival of the company, and sustainability is good business. The company is committed to better lives today and tomorrow, better business returns, and benefits for everyone, including economics and the environment. Customers want better performance, improved safety and less waste. This means prioritising efforts, dealing with waste management, distribution, environmental concerns and recycling.

Procter and Gamble addresses these concerns through an environmental management framework, which deals with safety, resources, compliance and social concerns. Global issues such as climate change are dealt with at the highest level, with expertise in risk management. The quality of data is vital, together with structures for organisation and networking. Within the detergents industry there is a code of practice. Procter and Gamble have set public targets, for reductions of 5% in energy, 10% in packaging, 10% in product and 10% non-biodegradable content, and these targets are monitored.

It is good business to be a good neighbour.

Discussion

Hans van Weenen noted that even large organisations can admit to their vulnerability. **Mary O'hEocha** noted the importance of customer and supply chain management, and pressures for compliance. She asked for a comparison of such pressures on SMEs in, for example, Sweden and the UK. **Erik Asplund** argued that pressure comes from both suppliers and purchasers, and that EMS are increasingly important. **Hans van Weenen** observed that company prescriptions could work, but would like research on when they are used. **Lena Weller** reported that in her experience of SMEs it was very hard to persuade people to attend seminars. **Axel Wannag** asked about the life cycles of SMEs, for whom long term planning was often an unrealistic luxury. **Erik Asplund** maintained that it is important for SMEs to have access to the information they need to make decisions. **Boguslaw Baranski** reported that many large companies require SMEs to have EMS programmes, and offer a 10% premium for environmentally clean products. **Ole Busch** highlighted the unrealistic objective of integration of policies across Europe. **Gerard Zwartsloot** observed that the normal sequence of priorities is quality, then environment, then OHS. It is still hard to gain market advantage from OHS, but there is local motivation that brings business benefits. **Ingvar Söderström** emphasised the role of the supply chain.

Andrew McCabe
Managing Sustainable Development at Shell International

The Shell Group is comprised of a large number of separate companies, and is not centrally controlled. Across the world the ethos has varied from "Trust me", to "Tell me" and "Show me". As trust diminishes, the need for transparency and involvement increases. **Andrew McCabe** reflected on Shell's problems with Nigeria and the Brent Spar. These matters are taken extremely seriously, with top priority for

Health, Safety and Environment board level and external HSE reports, which are externally audited.

Shell likes to see itself as a values based organisation, based on honesty, integrity, and respect for people. It has developed a statement of general business principles, which includes elements concerning the community and sustainable development. There is a clear Health, Safety and Environment policy: there should not have to be a choice between profits and principles. By the end of 1999 each Shell company will have an HSE management system, with environmental certification addressed by the end of 2000. This means a systematic approach, with targets, monitoring of the performance of contractors and joint venture partners, and an involvement of HSE in appraisal processes. It would help if there were certifiable OHS standards.

The approach is based on common sense, linked to senior management leadership and commitment. It involves hazards and effects management, product steward-ship, and sustainable development in the environmental, social and economic senses. The business case is straightforward, and reinforced by the fact that HSE systems provide precedents for good general business practice.

He ended by presenting a sustainable development roadmap and a sustainable development management framework, evidence of a holistic approach. Health, Safety and Environmental issues are considered together, both for thought and action, in a process which involves stakeholders and external reporting.

Hans van Weenen was glad to see evidence that the social component of business is being addressed at board level in major companies.

Ann-Beth Antonsson
Large and Small Companies

Large and small companies were compared. In large companies EMS could be well developed, but it was not clear how whole organisations could work with the system. What would be the driving force for integration of OHS with other systems, as opposed to separation? All too often there are different perceptions at board level and in the workplace. In small companies there can be external bureaucratic pres-sure to apply EMS. Integration is more likely, as SMEs are likely to use at most one system. There is pressure to minimise requirements.

Is the workshop concerned with the few who are currently working with Manage-ment Systems, or with the overwhelming majority who are not? Different strategies are required for the two target audiences. **Ann-Beth Antonsson** argued that systems do not necessarily provide knowledge, which is essential for successful manage-ment. Large companies have access to resources, knowledge and bureaucracy, and face the risk of major losses from loss of good will. Customers are an important source of pressure.

Small companies lack time, resources and knowledge; they are influenced by large companies, and are not bureaucratic in their processes. We need a heterogeneous model, encompassing many strategies, an emphasis on understanding differences, efforts to raise awareness of OHS and EMS and of the many strategies available for integration. Instead of seeking a single model, we should identify the core requirements.

Discussion

Hans van Weenen agreed that systems are procedures or structures, rather than sources of knowledge. He envisaged a patchwork approach, taking account of the uncertainty of risk in areas such as chemicals, which can present particular problems for SMEs. **Gerard Zwartsloot** argued that an effective EMS would address the use of chemicals, continuously reducing the use of untested items.

Wout Buitelaar declared that the workshop had been discussing old-fashioned management concepts, rather than modern approaches to supportive and participative management. In the expertocratic discussion there had been little mention of workers, who are stakeholders and tend to find the dangerous substances. There is struggle at the workplace level, and the location of responsibility is unclear. Traditionally the environment has been viewed as a strategic issue, while health and safety has been discussed at works councils. The alternative is for the tacit knowledge of the workers to meet the expert knowledge of the management, for mutual benefit. This is a route to eco-efficiency, taking account of the social dynamics of the firm, and defining quality in a way that includes the environment, working life and the product.

Ann-Beth Antonsson noted that ISO 14001 includes a requirement for participation, as does the Swedish and Norwegian approach to internal control. However, the emphasis can be on documents for the labour inspectors, rather than on real participation. In general, Management Systems are less effective when imposed top-down, and better when workers participate throughout the process.

Ole Busch reported on worker involvement in Denmark. EMAS does not require participation, but the European Trades Union Congress is pressing for change. In Denmark there is a requirement to consult the workers. **Michael Jorgenson** noted that the integration is achieved in Denmark through workplace assessments. Companies tend to continue in their traditional ways unless obliged to change, and stronger standards are needed to change cultures. **Gerard Zwartsloot** argued that standards require bureaucracy, but the social dynamics are just as important. **Hans van Weenen** asked whether EMAS and ISO 14001 can be described as democratic in origin and practice.

Stefano Farolfi reported from the food industry in Southern France, where systems have been externally required, but have not really been implemented within organisations beyond the minimum. There are issues of cost, knowledge and culture. **Jaap Bos** described research on the costs and benefits of voluntary systems: the economic benefits cannot be quantified with any precision, but they can be felt.

Gerard Zwetsloot reflected that the smaller the company, the higher the costs per worker of good practice in this field. **Jaap Bos** responded that the system needs to be adjusted to the company, and not simply vice versa. **Lena Weller** argued that we should avoid discussing simply costs, and should highlight the benefits derived from participation. Informing workers is not the same as consultation and participation.

Ole Busch favours moving forward with improvements to EMAS, making use of success stories, dealing with involvement and training of workers. He envisaged a role for the Foundation. **Margareta Mårtensson** declared that the draft for the revised EMAS does include involvement, and needs to be seen as a tool for both company and employees.

Boguslaw Baranski
International Guidelines on Good Practice in Health,
Environmental and Safety Management

Boguslaw Baranski described the process of work by the World Health Organisation in this field, and argued that it would not be possible to produce one set of guidelines or best practice applicable in each country. It is a question of identifying good practice to assist in policy development. A policy document is under development to be endorsed by the Third Ministerial Conference on Environment and Health in London in June 1999. The working assumption is that laws and institutions continue unchanged, that experience of quality management is available, and that industry is willing to participate. The outcome should be a safer, healthier environment and set of products. Good practice requires continuous improvement of HES systems, direct participation of working communities (where employers, managers and workers cooperate), the involvement of experts and strong community links. Core principles include legal compliance, continuous improvement, participation, the precautionary principle, community consultation, democracy and a preventative approach.

HES management in enterprises will not be solved by a particular standard, but by advocacy of a process, whereby targets are set by the working community, with expert support, verification through audit, and open communication. The process is based on the measurement of the impact of work on Health, Environment and Safety, using the environmental health and safety of the workers as indicators, and considering the broader context of occupational, environmental, health and lifestyle issues. Risks must be assessed and good practice adopted. This leads to the concept of "workplace preventable diseases", and requires underpinning by legislation and voluntary action.

The technical content of good practice in HES management in enterprises will comprise OHS, workplace health promotion, environmental health, environmental management, health and social capital, and quality management procedures. The economic appraisal will show the importance of Health and Environmental management, developing a tool to support decision making at corporate and national level. The document will be intended to support or initiate policy commitments, national objectives, agreed indicators and auditing of national systems. This requires the provision of external funding and knowledge.

The range of stakeholders is broad, and all must be involved, including governments (typically several different ministries), agencies, policy makers, employers, employees, NGOs, financial institutions, professionals and trainers. Many will be consulted through their governments.

Conclusions

Hans van Weenen concluded the first day with a brief summary of the key points made by each speaker. **Gerard Zwetsloot** had considered OHS in the context of QM and EM. **Jan de Saedeleer** of Procter and Gamble had addressed social concerns and the business to business dimension. **Erik Asplund** had highlighted issues for SMEs. **Andrew McCabe** had emphasised certification, transparency and top level support.

Ann-Beth Antonsson had asked who we are addressing, and noted that systems do not provide knowledge or guidance. **Lena Weller** felt that little was really happening, while **Boguslaw Baranski** had given an account of integration at governmental level.

There has been talk of continuous improvement, but of what? We can point to pioneering sustainable SMEs, and the greening of industry. **Hans van Weenen** introduced sustainable business development, production and consumption, and co-sustainable businesses based on partnership.

Second Day

Hans van Weenen introduced the second day, with a focus on practice, thinking about the future. What are the tools intended to achieve?

Participants were asked to correct the report of the first day of the workshop, and to complete a questionnaire, concerning recommendations on environmental management, health and safety, and suggestions for the Work Life 2000 conference.

He sought to stimulate debate on environmental management systems, suggesting that the adoption of EMAS and ISO 14001 was not enough for sustainability. The question of social impact remains to be addressed, concerning the organisation's ability to learn and adapt. There is a real sense of urgency, requiring immediate shifts of policy and practice. Prevention, urgency, harmony, participation, community and spirituality are key themes. He introduced a case of a German kitchen company, using natural substances and customer involvement in design. Wind energy is used, and delivery beyond short distances is not provided. The whole approach is integrated and holistic, and does not depend on an environmental management system.

Gerard Zwetsloot
Introduction to Case Studies

The case studies are also described in the book *Design for Sustainable Development*, provided for participants. They were collected from different regions, reflecting different cultures and strategies. Where possible, small or medium sized companies were chosen. Typically the examples were part of multinational companies, or part of networks. Allers is a small company example of 160 people.

A general format was chosen, dealing with an introduction to the organisation, aims and strategies, implementation, monitoring and control, evaluation and organisational learning, communication and cooperation, and notable findings. This is related to the plan–do–check-action cycle of ISO 14001, but with more attention to processes of organisational learning and the organisation of improvement processes.

Lena Weller and Dieter Kropp
Axel Springer Publishing, Germany

The work concentrated on an offset printing site in Ahrensburg. **Dieter Kropp** works on training programmes with the trades unions. **Lena Weller** is a consultant. Their study dealt with EMAS in the first 323 sites. The concern was for:

Motivation for Implementation

There were three major incentives: the pressure of public opinion regarding deforestation, pressure from the mother company (pressed by Greenpeace to develop a pilot site) and a local necessity arising from work with environmentally sound solvents.

Characteristics of Environmental Management System

The works council was integrated into the EMS, and was based on traditional OHS structures with added EMS functions.

Participation in the Process

There is no direct worker participation at Axel Springer, but a representative approach through works councils, and active information dissemination.

Roles of Representatives

The works councils act as initiators and trainers. Drawing on health and safety experience, they are working in project groups, and are closely involved in implementation.

Notable Findings

Elements for success can be defined, with a combination of external pressure and internal cultural development. There need to be working decision making structures, and working groups. Employee representatives need to be active in the environmental area. Effective worker participation is necessary at all stages, with real incentives. Underpinning this there needs to be training and information provision. Worker representatives need to see environmental protection as a task in its own right, as well as a strategic task in defending workers. Environmental management should add value to occupational safety and health work. Representatives need to see themselves as co-managers and moderators, part of a campaign for sustainable development.

Hans van Weenen recalled participation as a builder of a sustainable house. The timing and nature of participation are important. Consultation after decisions have been made is not full participation. **Lena Weller** responded that workers should be involved from the start. This enables them to act as promoters. **Boguslaw Baranski** asked about criteria for indicators at this site. **Lena Weller** had concentrated on communication, which **Boguslaw Baranski** saw as an indicator in itself. **Ingvar Söderström** asked about training as a preparation for participation. **Lena Weller** also saw training as a means whereby the works council participation was developed. The unions have no place in the company in Germany, or in occupational health and safety: this is the area for works councils. **Konstantinos Evangelinos** asked about the motivation for involvement of works councils in health and safety. **Dieter Kropp** responded that the works council sought to involve workers in health and safety training, with funding for training units from the company and employers. Organic solvents were a particular motivating area.

Jaap Bos and Gerard Rutten
Allers Bedrifswagens, Netherlands

Gerard Rutten set the case study in the context of the company with six sites across Northern Europe, with a total of 160 employees. The business is in truck sales and truck repairs, with parts stores and paint shops. It started as a family company, and has expanded and modernised. Quality has always been important, with the aim of keeping the customer satisfied and paying the bills. The working atmosphere must be right, with good housekeeping. The customer needs to have a good impression, and mechanics need the right instructions. Staff must give mechanics the right equipment and training. Motivation is important. Procedures, check lists and meetings are important: human factors including listening to people. Nonconformity must be reported and checked, and standard check lists are used for maintenance. Translated material is required for Eastern European workers and customers. Energy use needs to be monitored. Used vehicles need to be checked for leakage and PH levels. It was concluded that certain PH levels derive from the effects of acid rain: the subjects of environment and occupational health and safety overlap. Every week the manager identifies safety issues for discussion, and considers cases of structural failure. Accidents and near-accidents are recorded. There is a need to improve practice. This is where ISO standards can be helpful, enabling the company to show the customer and the government that standards are being met. Accidents have served to raise awareness of the issues. There is no standard for OHS certification. Overall the company works in a culture of Total Quality Management, with a process approach.

Jaap Bos considered reasons for implementing management systems. The process of dealing with trucks involves aspects of OHS, environment and quality in determining success. The stakeholders have different emphases. Clients, government and the public are interested in OHS (and the employers have to pay when employees are off work). Government, clients and the public are concerned with environmental issues. Quality is a concern of the Ministry of Defence and of clients, meaning that management systems conforming to ISO standards must be in place. The conclusion is that stakeholder views facilitate the synergy. Quality leads, and the others benefit, with links between Quality and OHS, environment and OHS, at times leading to conflict, but stimulating innovation.

Allers does not yet have EMS, but is concerned with sustainable development? Human factors are important, and new environmental technology is scarce. There is real commitment of management to sustainable development.

Boguslaw Baranski asked about training to avoid diseases that are not work related but cause sickness absence. **Jaap Bos** noted the activity of insurance companies, helping small companies survive. **Gerard Rutten** noted that sport can lead to injuries. **Mary O'hEocha** asked about data on health and safety, quality and environment conflicts. What can be measured? Quantification should be useful. **Ann-Beth Antonsson** described evaluations of the impacts of interventions. Cleaner technology means changing the working environment, and this can mean new problems in other aspects of the working environment. Measuring one aspect can be misleading, and a more holistic approach is needed. **Jaap Bos** agreed with a holistic view. Problems need to be discovered as early as possible.

Jakob Lagercrantz
AdeKemaAB, Sweden

This is a micro company, located in Western Sweden, producing soaps and car wash products. **Jakob Lagercrantz** is an environmental consultant and a board member of the company, having been chairman of Greenpeace Sweden. Sustainable companies need to make money in order to make sustainable products.

Who influences environmental change? Corporations were passive in the 1960s and 1970s. The Stockholm conference in 1970 raised political awareness. In the 1980s environmental issues were prominent, and in the 1990s some corporations are driving the agenda. In 1988 something changed the agenda: consumers started to care. Dioxin traces were found in coffee filters, through the process of chlorine bleaching. Consumers began to ask questions. The small company saw a niche for eco-friendly, eco-label detergents, and retail stores responded favourably. The small company made a strategic decision at the right time. Over the last five years all products have been reviewed to increase credibility. Ninety-six per cent of products are now environmentally labelled. The company has grown from 14 to 35 employees. They became organised, taking on ISO 9002, then BS 7750, ISO 14001 and EMAS. It represented a big commitment, but it was cost-effective for the company. In turn this raised questions of internal occupational health and safety. All staff were represented in the process.

The environmental work is more important than certification in itself. The initial environmental assessment made it possible to prioritise the environmental work, exploring opportunities and extracting strategic recommendations for the future. Development continues after certification. The company wanted to move ahead of legislation, facilitating recycling of plastics. If you are ahead, you can make money. The debate on PVC is long-standing: Electrolux and AdeKema decided to remove PVC, and communicated this publicly. An environmental report has been circulated, meeting EMAS requirements but also readable.

The results can be increased profits, an improved image, better contacts with authorities, and proud employees. Environment is a strategic decision. You don't have to do everything, but choose the appropriate tools. Keep developing, and keep it simple.

There are dangers in trying to do too much. It must be sustainable, avoiding dangers of bureaucracy. Credible communication means telling the truth, not "greenwashing".

Hans van Weenen saw advantages in being a micro company. Clarity is helpful, enabling a return to principles.

Raj Lakha
Britannia Refined Metals Ltd, UK

He gave an outline of the company, BRM, based in Gravesend. It is in non-stop production, with 400 staff. The business is metal refinery, lead refinery, in particular. Raw material comes largely from Australia. The lead is 99.99% pure, with other internationally renowned products. There are Safety, Health and Environment implications. There is a SHE philosophy of standards and openness. The company has a good reputation and a near monopoly position.

He provided a demonstration of BS 8800, a safety, health and environment standard. This is an integrated approach, and two models are offered, starting from either safety and health or the environment. BS 8800 is non-certifiable: this has proved successful among small companies. There are 7 core stages:

1. External and internal factors: e.g. the impact of the Asian crisis on metal exchanges, and the political implications of local government changes, demographic changes; financial, sales and marketing issues in the company

2. Initial SHE policy review, updated quarterly at BRM

3. SHE policy: commitment, organisation and implementation

4. Adjust policy, establishing a SHE culture: communication, competence, control and cooperation. At BRM this means regular communication, with weekly reports. There is a strong trade union presence, and trade union involvement in safety issues. BRM take their employees around the world, training them in modern approaches. SHE management involves everyone.

5. SHE planning: at tactical, operational and strategic levels

6. SHE performance measurement: active and reactive

7. SHE audit: environmental impact assessment and safety audits

BRM had the collective motivation to develop good SHE systems. BS 8800 has the advantage of being short and simple. It is sequential, with each stage documented. However, it can be limiting: how do we internalise external issues? It fails to take account of social relations in the organisation. Marx would be depressed! Insufficient account is taken of the ownership of capital, or of stakeholders.

In conclusion, he saw four key points.

1. Banks should take SHE seriously

2. Insurance companies need to be involved, and could insist on risk assessments

3. SHE management has not been seen as important: it needs to be exciting and prominent

4. Education and safety are important: BRM would like to develop an MBA course

Discussion on Cases Presented

Hans van Weenen noted that there are numerous preliminary steps before companies embark on the first stage in a process such as was described in the BRM case. He described sustainable SMEs, which are not characterised by environmental performance, environmental optimisation, or even eco-efficiency. Rather, we see values; vision and mission; quality of life and work; democracy; participation; solidarity; equity and equality; reciprocity; and a concern for the future.

Round Table Discussion

The core questions were:

- What do you think should be achieved towards sustainable production and consumption?

- Will environmental management systems lead to a sustainable future?

Margareta Mårtensson declared that there is no such thing as a standard company. All people concerned must be involved in sustainable production. She reflected on the next steps, with standards as applied to management. Problems are not likely to be solved simply through developing and issuing more standards. Every company has its own structure and culture, and strict standards may be inappropriate: this may inhibit involvement and participation. There is no standard view of paradise, nor is there an agreed sustainable future.

Ann-Beth Antonsson does not believe that management systems will solve the problems, as they do not provide knowledge. The main need is to improve knowledge, especially in SMEs. This will not be done through conventional university routes. We need other approaches to acquire good practice. The challenge is sustainable development, and today's good practice is not enough. We need simpler systems, more suited to the needs of SMEs, enabling benefits to be derived.

Wout Buitelaar likes bureaucracy in general. However, there is a difference between the structures involved in creating, developing, organising and transferring knowledge. Japanese work has dealt with these issues in the context of knowledge creation and transfer. There needs to be a strong organisation concept in order to develop a management system. However, the organisation is not a closed system. We need to consider customers and suppliers. Firms are dependent on each other, and inter-twined in processes of change. We need to deal with this inside-outside approach.

He then considered sustainable employment. Employment is increasingly with sub-contractors, working alongside others with different employers. People are uncertain, and are being asked to cooperate in these complex environmental areas. There needs to be some kind of certainty or security within the firm, if we are to collaborate externally in these new ways.

Axel Wannag was concerned with the soul: there had been accounts of values in particular cases, but we have been dealing with only 0.9% of European firms. The issues need to be raised at a national level. Government has to provide enough negative incentives to facilitate the scaling up of activity. He quoted Sir Thomas Browne (1690) concerning passion, and reflected that too many today are short-sighted, greedy and exploitative. As the Millennium approaches, he recalled "Where your treasure is, there your heart is also". So, where do we find negative incentives? This can be through internal auditing, transparency to the public, and business competition. Apart from the few enlightened cases, the method to extend good practice must include negative incentives. Looking into our souls, why do we have to continue the current state of affairs? He commended the Work Life 2000 initiative as a context in which these issues can be addressed.

Hans van Weenen opened the discussion. **Svend Jensen** argued for an enlightened firm, and discussed the level of consciousness, introducing transcendental meditation as a resource. The approach is popular in Japan and elsewhere, with considerable benefits.

Hans van Weenen then sought a collective answer. **Margareta Mårtensson** asked about the application of laws around the world. **Axel Wannag** noted that there is a mass of Norwegian law, but there are higher penalties for external rather than internal transgressions. There need to be real sanctions.

Boguslaw Baranski saw the need for a variety of tools, including education, legislation, and employee involvement. There could be no one solution, as each context, country and culture is different. There is a role for external consultants in addition to statutory services. Prevention of non-occupational diseases is also a key issue, yet we do not have a proper language in which to communicate about these matters. Universities neglect environmental health.

Sean Coleman spoke from a background in insurance, working with companies on health and safety. There is a problem that only a few companies are as yet involved. European VDU regulations are ignored because people do not understand why they are important. Education through case studies is important. In the UK HSE have looked at the costs of risk, and the insurance arguments can be compelling for managers. Security is a key issue for employees. Once "why" is understood, then "how" may be through a standard.

Ann-Beth Antonsson considered the chain approach, using links between companies. In Sweden and Norway SMEs move to internal control and environmental management systems as a result of external contacts, such as with suppliers and customers. **Wout Buitelaar** spoke about cannibalisation of SMEs in Japan, and considered different patterns of relationships in Europe. He traced the stages of environmental activity by firms, and argued for looking at levels below EMAS. The early movement from a reactive to an offensive approach, and then to innovation, needs to be understood. It is a continuing process, with a continuum of relationships. Complexity is a challenge.

Hans van Weenen favoured the comparative approach, and included consideration of SMEs in India. He noted the presence of the social partners, but asked about academics, NGOs and electronic networks. There are numerous active networks, which should become involved in our activities. This is a challenge for the European Foundation.

Ingvar Söderström
Conclusions and Summary of Proceedings

Ingvar Söderström started by explaining the nature of the Swedish National Industrial Safety Council, which is owned and funded by the Labour Market parties. The focus was at first on occupational health and safety. There is close partnership with the National Board for Occupational Health and the National Institute for Working Life, and links with every union and company. He liked the practical approach that had been taken in the workshop. If you do not understand your history, you cannot address future problems.

Can we make better links between the external environment and the working environment, which need to be handled together? We have had accounts of integration with the objective of sustainable development. There is a gap between external and internal environments. There is different legislation, driven by different authority structures, in each country. There need to be new forms of partnership.

Environmental management has been an issue for managers, while the working environment has involved employees. It is popular to talk about SMEs, but we do not necessarily have the same definitions. If the defining level is 500 employees: dry cleaning shops are much smaller. Many SMEs are in fact big companies, but will never have large numbers of employees. Farms are typically run by SMEs, but have vast potential in the environmental field. An investigation of communication among Swedish farmers showed high levels of use of PCs and the Internet. The internal and external environments are brought together. There will still be differences according to size, in whatever country. SME managers are more likely to be craftsmen than businessmen: they are sceptical about laws and regulations, worried about external inspection. Every change has an impact, often additional costs. By contrast, big companies have enormous personnel resources with which to deal with these issues. We probably need some kind of service support for the smallest and poorest companies. We cannot separate environmental questions from work environment. Large and small companies are worried about media coverage: this causes external and internal problems. Employees are becoming more active, with trade union programmes developing.

The environmental management process cannot be forced by legislation or standards: it is a learning process. It is based in the corporate and country culture and traditions. We must take this into account. The changes take time.

How, then, can these issues be integrated in competence development in companies? It is not just an issue for employers. The labour market partners have opportunities, and there can be a role for academics.

The Swedish King had opened the ICOH conference in Stockholm with an emphasis on sustainable development. The Swedish Prime Minister now places the issue at the top of the agenda, with unemployment. It will be a priority concern for the Work Life 2000 Conference in Malmö in January 2001.

Ingvar Söderström ended with thanks to the European Foundation, to **Hans van Weenen** as chairman, and to all participants.

Workshop Participants

Ann-Beth Antonsson, IVL, Sweden
Erik Asplund, NUTEK, Sweden
Boguslaw Baranski, WHO
Jaap Bos, NIA-TNO, Netherlands
Wout Buitelaar, University of Amsterdam, Netherlands
Aidan Burke, Construction Industry Federation, Ireland
Ole Busch, consultant, Denmark
Sean Coleman, loss prevention consultant, Ireland
Torsten Dahlin, Svensk Industridesign, Sweden
Jan de Saedeleer, Procter and Gamble Europe, Belgium
Luigi Doria, Bocconi University, Milan, Italy
Richard Ennals, Kingston University, UK
Birgit Erngren, NUTEK, Sweden
Konstantinos Evangelinos, Imperial College, London, UK
Stefano Farolfi, ENSA, France
Sarah Farell, European Foundation
Anton Geyer, PPM, Austria
Lars Grönkvist, NUTEK, Sweden
Geoff Hayhurst, Northern Technologies, UK

Arne Helgesen, HK Denmark, Denmark
John Hurley, European Foundation
Svend Jensen, Naturlovspartiet, Denmark
Anders Jeppson, Arbetsmarknadsdepartementet, Sweden
Michael Sogaard Jorgenson, Technical University of Denmark
Dieter Kropp, IQ Consult, Germany
Jacob Lagercrantz, Ecoplan, Sweden
Elisabeth Lagerlöf, NIVA, Finland
Raj Lakha, Safety Solutions, UK
Jan Grolier Larsen, Landsforeningen Levende Hav, Denmark
Bernard le Marchand, FEMGE Belgium
Margarita Lazcano Nunez, National Institute for Occupational Safety and Health, Spain
Henrik Litske, European Foundation, Dublin
Maria Mandaraka, National Technical University of Athens, Greece
Margareta Mårtensson, SAF, Sweden
Andrew McCabe, Shell, Netherlands
Joe Mongan, Tullow Oil, Ireland
Mary O'hEocha, University of Birmingham, UK
Filomena Oliveira, European Foundation
Clive Purkiss, European Foundation
Irene Reuther, SKZ, Germany
Cheryl Rodgers, University of Portsmouth, UK
Gerard Rutten, Allers Bedrifswagens, Netherlands
Jens Peter Schytte, Skolen for Okologisk Afsaetning, Denmark
Antonio Scipioni, University of Padua, Italy
Lena Skiöld, NIWL, Sweden
Ingvar Söderström, Swedish National Industrial Safety Council
Christina Theochari, Department for the Environment, Greece
Carl Thornley, Northern Technologies, UK
Efi Valiantza, Cleaner Production Centre for Greek SMEs, Greece
Hans van Weenen, IDEA, Netherlands
Peder Venge, Project secretary, Aarhus, Denmark
Eric Verbogh, European Foundation, Dublin
Axel Wannag, Labour Inspection, Norway
Lena Weller, IQ Consult, Germany
Gerard Zwetsloot, NIA-TNO, Netherlands

Reflections on the Workshop

Hosted by the European Foundation for the Improvement of Living and Working Conditions, the workshop gave strong support for an agenda of sustainable development, and built on a pre-existing programme of research and publications. The workshop made effective use of the two days available, using published case studies linked to the keynote presentations. It was a truly European event, enabling a number of different national and European bodies to work together, and offering the potential for coordination of future plans. The theme of sustainable development may find a place in the remaining stages of Work Life 2000.

4. Occupational Trauma: Measurement, Intervention and Control

The workshop was led by Tore J. Larsson and held in Brussels, 14–16 December, at the Offices of the Swedish Trade Unions.

Tore J. Larsson
Presentation of Themes, Keynotes and Aims for the Workshop

Tore J. Larsson introduced **Arne Wennberg**, Secretary-General of Work Life 2000, who gave an overview of the Work Life 2000 conference and preparatory workshops.

Tore J. Larsson, originally Swedish and based at Monash University, Australia, then introduced the three keynote speakers, who offer three different approaches and should enable the workshop to engage in lateral discussions. **Tord Kjellström**, originally Swedish, is Professor of Environmental Health at Auckland University in New Zealand, and formerly worked for the World Health Organisation. **Paavo Kivisto**, originally Finnish, is Deputy Minister of Labour in Ontario, Canada. **Harry Shannon**, originally British, is at McMaster University in Canada, where he is Professor in the Department of Epidemiology and Biostatistics, and is a member of the team that has written a manual on safety interventions at work for ICOH.

The workshop was intended to be open, driven each day by a keynote speaker and with groupwork and presentations of country cases. **Tore J. Larsson** emphasised the pace of change in industrial structures relevant to risk, with economic and political dimensions.

Tord Kjellström
Approaches to Quantifying the Public Health Impact of Occupational Trauma

Tord Kjellström had conducted research on cadmium poisoning before engaging in environmental health and statistical work in Sydney, and in Auckland. He argued for proper weighting and ranking of occupational traumas, which requires the collection of statistics within an overall framework. It could be argued that in light of medical advances occupational trauma is of increasing significance in Europe and the West, by contrast to Africa and the developing world, where the dominant new factor is AIDS.

The study should be in terms of the work environment, and not just worker behaviour. Somehow we need to be able to record details of incidents that did not quite cause injury, and we need to see the relationship to public health. We need to be able to highlight the fact that the preventable fraction of injuries is of considerable importance, justifying intervention. Politicians and decision-makers find such information difficult. Looking at figures on incidence and fatality rates in the EU, he noted that as workers get older, the rate of accidents declines, but the rate of fatalities increases.

The "big picture" concerns communication, packaging information so that it can stimulate policy decisions. We need to develop ideas of health expectancy, and to consider new patterns of argument, for example based on discounted life years, in order to clarify, for decision makers, the relative impacts of future events. One useful measure is of cumulative mortality risk between the ages of 15 and 60, for which figures are available, showing greater risks in Poland than in other countries represented at the workshop. We need to highlight the indirect consequences of occupational trauma: for example the effect on families, or the effect on small businesses of an accident involving the owner. Effects can be cumulative. There are also considerable displacement effects on finite health service resources. We should not neglect the fact that unemployment, a key priority for all European Union member states, is bad for the health: some may argue that it is better to be at work, even in a workplace whose safety could be improved. Perceptions of risk need to be clarified, with communication improved and oriented towards appropriate kinds of action.

Tord Kjellström sees occupational trauma as an issue of public health rather than industrial relations. He is happy to adopt a body count mentality in order to achieve impact, and argued the case for a "vision zero" approach, whereby successful work is seen as accident free, and thus any expenditure on improving safety is seen as an investment in successful work, in productivity. Given the right standards, and the means of showing benefits, this approach can be effective. It is a matter of moving on from the rhetoric of "acceptable risks", as put forward by the nuclear power industry, and in a way that impresses economists. We know that paradigms can shift, but that it takes time (examples included democracy, human rights and rights for women).

He conceded that it is harder to talk of exposure assessment for occupational trauma than for chemicals. It can be done: he cited the work of Volvo on the likelihood of car accidents at different speeds, recording the effects of safety belts. It is a matter of quantifying exposure levels.

Overall it is a matter of moving from data, to information, and then to knowledge. For this to be effective on an international scale we need to be able to harmonise practice and recording. The objective should be "evidence based" public health and trauma prevention.

Group Working

Participants then had conversations with their neighbours, before introducing them to the workshop. Two groups were formed, to discuss the quantification of Occupational Trauma data, and epidemiological reasons to question trends and comparisons.

How is it decided how to present data? How can we trust data presented by others? It clearly depends on the audience for the presentation. The workshop was concerned to target politicians and senior managers, deciding about resource allocation and priorities. It would be useful to have overall guidelines on data presentation, providing some quality assurance to avoid the worst kinds of "lying with statistics". Many decision makers, such as the Polish Labour Inspectorate, are not interested in epidemiology. Statistics are used from the local area, and these may not be connected with national epidemiological research.

It is important for data to be comparable, enabling learning from differences. If it is used in this way, the issue is of local area representation. Can you not use someone else's data instead of having to generate your own? Policy makers should be able to see the relevance of data for making decisions. There is often a gap between data and user, and a bridge is needed. Statistics should not be thought of as just "damned lies", but should be made easier to understand and evaluate. Epidemiological data should be of use in insurance, clarifying the nature of risk. Data needs to be understandable, but limited in quantity. Researchers need to develop methods of easy presentation. One is to locate the performance of the target organisation on a graph showing the performance of rivals. When accident rates were compared on the Swedish and Danish sides constructing the new bridge, differences were found and there was pressure for working practices to change.

The Netherlands lacks reliable accident data: does this mean that policies and approaches suffer as a result? How then are decisions made? In the case of natural disasters, there is general predictive data, but there is no reliable data on occupational trauma. The problem is left to companies, and studies are undertaken in particular industries, drawing on foreign data. Should certain data collection be required? It was argued that in general policy changes are evaluated by feel, and not in terms of rigorous data. There is more data available in Sweden and Finland, provided through the compensation system. However, as more burdens are passed to employers, less reporting takes place. This makes consistency of data almost impossible to maintain. Different systems will give different results, depending on whether data comes from labour inspectors, hospitals or compensation claims. Can we really use national figures of work related fatalities where exposure risks are not considered? When reductions in fatalities at work are announced, there is rarely discussion of the change of hazards, or of improvements in medicine, which may mean accidents cause disability rather than death.

What are the effects of epidemiological information on decision makers? It is a matter of engineering their perceptions. Comparisons with the average, or with benchmark cases, are most effective.

Examples of Public Health Impact

Tore J. Larsson had prepared three case studies, based on a farmer, a concreter and an apprentice, and these were the basis for group discussion. The farmer experiences high exposure to risk, but tends to neglect safety issues. Concreters are good examples of small sub-contractors, self-employed and tending to suffer physical deterioration. Apprentices are on the border between adolescence and adulthood, and often suffer abuse. All fall within areas of current political concern.

The cases were discussed by the groups, considering the need for quantified data and proposals for new ways of presenting data, including exposure, cumulative risk, and Public Health impact.

It is clear that there is consistent under-reporting of farm accidents in Sweden and Finland, partly due to the system of compensation. It is recognised that there are problems in Denmark, which were reduced through a combination of improved design, more information and industrial restructuring, halving the number of accidents. It is known that certain models of tractors tend to tip over, but they have not been banned. From the Netherlands lessons can be learned from samples of hospital data, but there is no link to occupational data. In Poland, Labour Inspectors make only periodic visits to farms, where they have no enforcement authority.

Comparisons are difficult. Even if we have accident surveys, farmers are often not included. How are they to be convinced of the importance of safety? The answer seems to be mutual inspection (as in Sweden) and group collaboration or networking, for example through farmers' associations. Many farmers are not interested in being safer: there is a macho culture, which emphasises that the work just needs to be done; if there are accidents, the community will help.

The picture is complicated by unofficial workers, and the growing "black" economy, avoiding overheads such as taxes and safety, with effects on all three cases. Laws are hard to enforce. Limited data is available from public health systems, and benefits often remain unclaimed. Harmonisation of compensation arrangements will not be easy. Provision of subsidies and income support is controversial. In the Netherlands employers are responsible for the first year of sickness absence, which further complicates reporting.

More information is needed from interviews and larger surveys. In each of the three cases, we must accept that safety is not always seen as paramount. What kind of data do we need to collect? Sometimes what is needed is action, including raising awareness of preventive measures. The underlying issue is of acceptable risk. Many training and placement programmes do not include provision for health and safety inspections, even where hazardous materials are used. In these cases, data provision can be used for empowerment.

In the present European political situation, employment is a priority, and safety arguments may not be heard. This is further complicated by changes in the nature of employment. Part-time and short-term working, in controlled and uncontrolled areas of work, have considerable implications for hazards and for labour inspection. The Robens principle was that work is work, no matter in what industry, but it did not fully tackle the boundary between employed and self-employed, or paid and unpaid, and health and safety criteria are relevant in each case.

Is it enough just to provide information about dangerous working conditions? Should there be some kind of compulsion, or a link to insurance? After all, the public health service incurs costs as a result of farm injuries. It was argued that many concreters do not get their money's worth: they pay insurance premiums but do not receive the attention given to larger customers. Imaginative approaches were discussed, including risk pooling and cooperative policies. The equivalent of Micro-Credit (as pioneered through the Gremeen Bank) should be available through insurance. It was recalled that insurance had been an original function of craft trade unions, which often developed into mutual insurance companies, many of which have since converted to operate on a commercial basis, preferring large customers. There were worries about information imbalances in insurance, to the detriment of the small business.

Insurance has become part of a distorted set of economic indices, in which the costs of clearing up major disasters, or repairs after accidents, are seen as part of Gross Domestic Product. Insurers may feel no need to be interested in prevention.

In terms of data presentation, it is helpful to focus on measurable items. In order to make sense of comparisons, we need to understand the nature of social back-up systems which complement compensation systems. We need to go beyond annual reports. Ideally we will identify a format that is shown to be readable, and set up systems to deal with ongoing data flows.

Some participants were sceptical about claims of a data shortage, and argued that the real need is to relate the data, and the research that produces it, to experience of daily life. The issue is the contrast between data aimed at producing action and data aimed at facilitating decisions. Lessons can be learned from environmental health, with the identification of risk situations and the clarification of links between consequences and precursors. This approach has been taken with small businesses in Ontario, using a variety of means of communication.

Tore J. Larsson referred to the argument that accidents are due to our behaviour, whereas disease can happen to anyone. This presents problems for macho males, and for an epidemiological approach to accidents. Accidents are not contagious or random. In borrowing ideas from epidemiology to apply to safety, it is not always clear how exposure can be quantified. Exposure can be close to or remote from the accident. All too often our interventions are about preventing the previous accident. **Kari Häkkinen** declared that we only understand epidemics after they have happened.

Occupational traumas may be apparently unrelated to the work in hand. **Tore J. Larsson** gave the example of Swedish firefighters who suffer football injuries when taking exercise between fighting fires, and concluded that only 0.5% of their time is spent on ostensibly dangerous work.

It was agreed that engineers should be taught to understand human needs. Countries should be better able to learn from each other, while at present each seems to engage in separate tests. Harmonisation and mutual learning are needed. One example was work on ladder safety in Finland, which has not been taken up elsewhere in the European Union.

Tore J. Larsson
Presentation of the Theme for Day 2

The three perspectives of the workshop were recalled. **Tord Kjellström** had introduced discussion of long-term consequences of trauma. Often we see trauma as different from disease, as a one-off event. The epidemiogical perspective is refreshingly different and thought-provoking.

The second keynote theme, from **Paavo Kivisto**, is of government approaches to occupational health and safety. **Tore J. Larsson** described a crisis in labour inspectorates, over policy, resourcing, staffing and political support. Many countries face difficulties in this area. He drew on experience of Europe and the Anglo-Saxon world. In Canada there has been a trade-off between free market, social control and welfare. British Columbia is interesting, and Ontario is another good example.

Paavo Kivisto
The Role of Government in Advancing Occupational Health and Safety

Paavo Kivisto is Assistant Deputy Minister of Labour in Ontario, responsible for operations. He set the agenda with a presentation before the workshop groups addressed questions. He described the progress in reforming Ontario's health and safety system, noting there is scope for improvement.

He dealt with government priorities, across the world, and the civil service context in Ontario. Outlining the role of government, he then considered performance measures, addressing customers and workplaces, and assessing outcomes. The government is accountable to the citizens, taxpayers and the clients it serves. A recent international survey suggested that governments are pressed to reduce deficits, improve productivity and quality, and to privatise. In Ontario this has meant focus on core business and ensuring quality service. A new mission statement stresses advancing safety, fairness and harmony. Where government need not undertake a particular role, it can be passed to others. How can inspection and enforcement be handled in future: what should be the structures and approaches within government? It means responding to citizens who want timely service, less red tape and a clustered single window service. The Ontario government is committed to removing the budget deficit by 2001, and thus far is on track.

The budget of the Ministry of Labour has been reduced significantly, resulting in less staff resources than in the past. The Ministry has stopped doing certain things. Free advice is no longer provided by government, with a shift in responsibility to the private sector. Laboratories have been closed, and services are contracted instead. Ministry hygiene, ergonomics and engineering resources are used to support internal inspection and investigation activities. Responsibilities are being assigned to the employer, who is held accountable. Common standards are maintained. Regulations and permits have been removed where possible. There is still reluctance by private sector engineers to take responsibility for health and safety. It requires a change in mentality, a transition strategy and cooperation from professional licensing associations. The government works with partners to fill the new gaps. Safety is no longer the responsibility of government: safety in the workplace is the responsibility of those in the workplace. The Workers Safety Insurance Board had been assigned the prevention mandate in addition to handling compensation claims. They are shifting their large staff from processing claims to addressing prevention. The Board also funds sector specific safety associations who provide information, training and health and safety expertise.

The role of government is to set the vision, develop public policy, and to manage the delivery of programmes and services, where they continue as public sector functions. In occupational safety the task is to encourage investment, jobs and prosperity. The best companies find quality fully compatible with health and safety. Ontario is striving to become among the safest in the world; with flexible and responsive labour laws. Government's role is setting, communicating and enforcing standards. Targets were set in terms of reducing lost time injury frequency. Strategy is based on working with the Board and safety associations, putting an effective prevention strategy in place. This means encouraging self-reliance, and focusing the Ministry on the task of setting and enforcing standards.

The safety, productivity and quality of poor performers have to improve, approaching the standards of the best, with cooperation at workplace level. High performing workplaces are used to provide advice and support. Providing recognition helps. Financial incentives and penalties via insurance can be effective motivators. Promotional campaigns are used to raise awareness and socially market health and safety, dealing with students and young workers, outreach to new business, and the promotion of self-reliance.

"Safe communities" is an example of a self-reliance initiative. It is private sector funded and driven, leading to community based planning and delivery. One key component targets small business. Banks and a few private sector companies were enlisted, investing seed money in safety initiatives, and leading to remarkable savings in compensation costs, and less marked reductions in injuries. The initiative is successful in targeting small businesses, which have long gaps between injuries and do not see health and safety as a problem. They are hard to contact, and it is impossible to provide regular visits.

"Safe Workplace Ontario" was based on sawmills, focusing on those seen as unsafe. Twenty-one key elements for a health and safety programme were agreed, and sawmills were invited to register, receiving free training, manuals, and policies, enabling them to improve their health and safety programme with an independent audit. This led to a 19% reduction in lost-time injuries.

The Ministry targets 18 sectors, allocating staff time according to injury frequencies. Poor performers are given attention. Information on injury rates and compliance is computerised, enabling comparisons and prioritisation of inspections. Enforcement activities are increasing. Prosecutions are considered following deaths and non-compliance with enforcement orders. Not much is changed by prosecution, but it generates publicity.

So what has been learned? It is important to have a clear vision and business plan. Stakeholders need to be on board. Transition plans are needed, which do not leave essential services undelivered. Clear measurable goals must be set, and met, tying individual staff contracts to the overall business and operational plans. It is important to see outcomes: in this case achieving fewer injuries and illnesses. High schools will be teaching health and safety. A number of parallel initiatives are needed.

Challenges include delivery of an effective injury prevention service, improved research, education, measurement, performance based standards, and emphasis on small business, and the continued restructuring of government operations.

Examples of Role of Government

Paavo Kivisto then introduced the questions to be addressed by the groups, starting with Inspection and Enforcement. He suggested challenges for the European Union, and for employers operating across borders.

Inspection

The need to enforce standards can arise when contraventions are found during investigations, or during planned inspections. The limited resources available to governments need to be deployed to have the maximum impact on enforcing compliance and reducing workplace injuries and illnesses. Inspections need to be targeted to workplaces that are at high risk of being non-compliant. The actual inspection strategy in the workplace has to be efficient in assessing the level of compliance. An inspection results in corrective action by the workplace parties for any contraventions, but more importantly, it should result in improvements in the workplace policies and programmes, to prevent future non-compliance and the need for further government intervention.

The discussion considered inspection and auditing visits, with an emphasis on the changing role of governments. Much depends on what roles have been taken on by others. Can the problem be addressed by certificating safety management systems, as is being tried in the Netherlands? What role does this leave for government? The danger is that government inspectors switch to checking what has been written, not the situation in the workplace. Danish inspectors are required to see the work environment. The use of auditing has been over-extended, with a focus on documentation. Strange accidents have occurred in audited workplaces. It was argued that effective auditing needs to go beyond the paperwork, looking at practice. Is the audit then against the company's paperwork, or against some externally defined system? We need competent inspectors and auditors, with experience and understanding of industrial risk. Expertise requirements vary. The inspector must be able to define critical risks.

There was discussion of the responsibility of the company board to shareholders, and of human resources balance sheets. Experience in environmental health was compared, where the profile of key issues for the future has been raised by groups such as Natural Step. There has been certification of environmental systems, in contrast to occupational health and safety management systems, where employers have resisted. There is a case for an ISO standard, but employers have resisted standards for Health and Safety management, which may be because they fear increased costs. Responsibility lies at board level. There has been no internal reporting mechanism, so board members are ignorant, and fear a straitjacket.

Why has there been more progress on environmental matters? It is a global issue, while health and safety is more internal. Quality marks can be popular with companies for promotion and marketing purposes. There has been resistance to ergonomic standards in Canada, but these are present in European norms. There could probably be no single standard, as cultures vary, and safety audits can focus on different issues. Inspectors and auditors have different roles. Inspectors have experience of enforcement, and may not have the necessary independence to be auditors. This may be a task for specialist companies.

Trauma risks are not normally seen in terms of environmental exposure. The criteria for environmental use can conflict with occupational health and safety, and there can be a "not invented here" phenomenon.

Kari Häkkinen described the Finnish ASKELMA system of safety assessment, looking at hazards, activities and accidents, developing action plans. This goes beyond inspection and auditing, and involves building a partnership with the company. The starting point is accident statistics, informing discussion of particular risks. Initiatives often come from companies, but the government holds lists of critical companies. Company systems can be at different levels of development. Andrew Hale asked whether different approaches are needed for companies at the different stages of development. Harry Shannon asked about the generality of the system, given the variety of needs. Ingrid Christensen talked about the differentiated approach taken in Denmark, with three identified types of company based on the will and capacity of the companies concerned. This appears to be a European trend.

Tord Kjellström asked if a new word is needed, replacing inspector or auditor. Andrew Hale asked what the role is: is it checking and possible prosecution, or is it improvement of the management system? Privatisation has changed practice and roles, often removing improvement as a government task, and moving direct to prosecution. **Tord Kjellström** argued for evaluation of the changes. **Kari Häkkinen** asked whether the roles can be combined in the same visits. **Ingrid Christensen** argued that inspectors must be able to present the case for investment in health and safety. **Andrew Hale** observed that Dutch inspectors are no longer allowed to give advice: many leave and become advisers, to be replaced by those with a policing mentality.

Tore J. Larsson reflected on the emerging new role of the inspector in Europe, and drew the parallel with the traffic policeman. **Paavo Kivisto** argued that the inspector needs to understand how the organisation works. He needs to reinforce what is right, as well as highlighting what is wrong. Investigative inspection in Ontario can be lengthy and thorough. **Tore J. Larsson** thought that this was an idealistic picture: responsibility for improvement is being given over to companies, while governments focus on targeting compliance. **Andrew Hale** asked whether it is the role of government to increase self-reliance. **Kari Häkkinen** noted the complexity of labour inspection compared with road traffic law. There need to be priorities set. **Tore J. Larsson** argued that all too often responsibility for health and safety has been delegated to the risk-exposed workers.

Enforcement

One of the roles of government is to enforce compliance with legislated standards. However, the criteria to define what an effective enforcement programme should be are not clear. Enforcement can be a successor to investigative inspection. It is sensible to target "black spots", but they can be hard to identify, and they change. Cultures vary between countries. The Finnish, Danish and Polish experiences are different, with Greece different again.

We talk of machine standards, and official assessments, but we know that many manufacturers do not follow the standards. How can we deal with this area of enforcement? We need new ways. Should responsibilities be publicly located? Should punishment be separated from advice? Cultural differences continue.

Why is enforcement needed, and in whose interests? This appears to be a typical Dutch question. We need to understand the foundations of regulations that are being changed. **Tore J. Larsson** gave the example of the North Sea, and the philosophical rationale for doing what is possible. Complex forms are not applicable to small businesses.

Paavo Kivisto described accidents, largely falls, in the construction industry. Inspectors communicated as hard as they could, seeking to avoid prosecutions, but enforcement proved critical in the absence of a mature safety culture. **Andrew Hale** argued that health and safety authorities should not be kind to everybody: some businesses are run by criminals. Inspectors still seem to spend time with the big companies (like the drunk under the streetlight, and because they get a better welcome).

Tore J. Larsson praised the Gallic dialectical approach of Ontario, combined with a North American sense of business. In Australia there is a view that enforcement is not needed. In Europe labour inspectorates may die out, with reduced funding and increased bureaucracy. There needs to be a flying squad of tough inspectors to catch the crooks, but also ongoing development work to assist in developing self-reliance.

Paavo Kivisto argued that a good workplace has good health and safety. Andrew Hale asked about linking health and safety databases with databases of other variables which might be predictive of poor Health and Safety management, such as financial variables and tax systems. Paavo Kivisto reported that a system is currently being launched. There may be a move to a single system, which may help target inspections. He noted that there are numerous competing agendas, especially when inspection systems are being changed.

Setting Standards

Government has sole responsibility for setting minimum standards through its legislation. There are several options and issues related to the nature of and alternatives to government regulation. What is the role of industry "best practice" and practical guides?

Some in industry prefer regulation, lacking confidence that codes of practice would succeed in bringing companies into line. Governments are inclined to pass from regulation to guidelines and standards. When should there be regulation? What form should it take? Should it be prescriptive or performance based?

The group had explored the legal arrangements in each country. How are inspectors able to use the standards in a similar way? How can companies be expected to use the standards consistently? Good methods and best practice can provide a basis for advice, while standards must be followed. This applies especially in high risk areas. Minimum standards can be useful, but it can be hard to go further. The Dutch position was that standards should only be created when they are needed. European directives are not necessarily implemented by all members. The problem was not solved. Increasingly standards are set at a European level, with the goal of harmonisation rather than safety and health. Tore J. Larsson argued that standards are frequently vague, corrupt, and influenced by industry interests. Kari Häkinnen argued that some standards have saved accidents. Andrew Hale maintained that you need to be rich to participate in standards development at European level. Tore J. Larsson gave details of fork-lift truck standards which are much too vague to be used as design standards.

Building Self-Reliance

By working with clients to increase their desire and ability to be self-reliant, government can shift resources to workplaces needing its intervention. Many jurisdictions have implemented innovative approaches that have been effective in getting employers to take more "ownership" in improving their health and safety performance. How can this approach be extended? What is the role of government? What techniques are available for others to use? What is the role of education?

The group explored the nature of self-reliance. Organisations should be able to take action, exercising internal responsibility. Each workplace should be able to solve its own work and safety problems without outside intervention. This has been the starting point of Finnish insurance. The emphasis varies across Europe. Andrew Hale said that creating a non-governmental external service, as in the Netherlands, does not necessarily increase self-reliance, just transfers reliance from government to the service. **Kari Häkkinen** noted that Finnish shipyards use off-shore inspectors, and have developed self-reliance.

Tore J. Larsson argued that any small business operator sees himself as self-reliant with respect to risk. Safety is seen entirely as individual behaviour. The biggest problem is one of attitude. **Ingrid Christensen** recommended linking health and safety with other factors. **Tore J. Larsson** argued that this does not help with sub-contractors. He is sceptical about the role of inspectors. **Inga-Lill Engkvist** talked about built-in risks, which may be missed by those in the workplace. **Tore J. Larsson's** farmer and concreter are self-reliant. **Kari Häkkinen** argued for a partnership approach. **Andrew Hale** argued that small businesses can be too self-reliant, and often need to understand other ways of doing things. **Tore J. Larsson** indicated that this is in effect an issue of prevention: we need to build networks at local and regional level, enabling businesses to talk to and learn from their colleagues.

At what level should we seek self-reliance? Should it be at the level of the individual firm, or through alliances? The French Insurance system has funding to support preventive activity for small businesses at regional level. This provides a valuable model for labour inspectors to follow. **Mette Dyhrberg** indicated that this represents a positive experience with risk-taking, but may not only be a matter for labour inspectors, but also a responsibility of the labour market parties. **Ingrid Christensen** highlighted the success of the ILO approach with small and medium sized businesses, with employers visiting each other. **Kari Häkkinen** noted that an inspector is now also expected to be an educator and facilitator as well as a technical expert. A police role prevents the facilitation role. **Andrew Hale** quoted Irish experience of large companies supporting smaller companies in health and safety. Inspectors have evaluated the improvements. Perhaps there are cultural dimensions, and different perceptions of good neighbours. The approach might not work in the Netherlands, but appears to do so in Norway. The key is the capacity to transfer knowledge, and take on some of the previous role of government advisers. **Tore J. Larsson** saw it as less likely to work in an adversarial culture. **Andrew Hale** confirmed that reports of UK work showed concerns about liabilities. The answer can be a contractual arrangement such as pioneered in the Netherlands with certification of contractors to be allowed to be on the bid list for jobs.

There was discussion about the applicability of the same approaches to risk in large and small companies. Some procedures and check-lists are inappropriate for small businesses or those whose work is mobile. Small documents could be produced for small businesses. There can be inexperience with accidents in larger companies, where it is necessary to make it credible that accidents could still occur. The story needs to relate to the experience of those in the workplace up to the point of the accident. Near-accidents are important.

Large companies tend to use documented systems, which need to be changed if they are to be transferred to small businesses. **Ingrid Christensen** described how materials are changed for the new audience. **Inga-Lill Engkvist** argued that people need to understand about risk, and to be prepared to prevent them.

How do you determine the competence of the small business? Is it demonstrated by what goes on? What tools can be used? It is perhaps a matter of dialogue. This seems somewhat *ad hoc*. It may not be appropriate to ask for documented systems. **Ingrid Christensen** argued that the trained expert knows what to say. It cannot be guaranteed that different experts come to the same conclusions, but engaging in exercises with inspectors can help. There might be focus on a particular type of injury, such as to the back.

Accidents are derived from root causes, not from the general environment. How can an inspection programme address this? The root causes concern management, organisation, technology and processes, communication and culture. It is a matter of understanding the whole process, behind the surface of unsafe acts and conditions, then accidents and injuries. A self-reliant organisation has feedback loops, and can respond. The models are not new, but are ignored by those who do not think about safety every day. We need a systemic approach, probing beneath the surface to the culture, and not just reacting after the event. **Inga-Lill Engkvist** reported the positive effects of feedback to the work group. One problem is to elicit the norms and tacit knowledge of the particular workplace.

Andrew Hale restated the problem of determining the right level for self-reliance. For governments, it can just be a matter of passing the responsibilities on to others.

Tore J. Larsson
Presentation of the Theme for Day 3

The day was intended to present the third piece of the jigsaw. Epidemiology was the first theme, as a means of improving motivational propaganda. The second theme was the role of government, based on an account of the government of Ontario. The final theme concerns appropriate interventions to reduce risk. There has often been a lack of scientific method in what is an important area. **Harry Shannon** has been a leading figure, researching and editing a manual for safety interventions. The presentation would lead to discussion of examples.

Harry Shannon
Successful and Unsuccessful Interventions to Prevent Occupational Trauma

Harry Shannon is Senior Scientist at the Institute for Work and Health, as well as at McMaster University. The research concerns the workplace, workforce and rehabilitation. Many interventions are based on studies of correlates of safe performance. There are a number of types of evidence, starting with work on individual characteristics such as "accident proneness". It is hard to separate individual "proneness" from the risks of the job. There can be studies of human error, and behavioural research. Studies of technical issues can highlight machine guarding, risk analysis, materials strength, and objects involved and biomechanical analyses. Identification of "causes" and "patterns" of accidents can involve modelling of processes. Depending on the investigator, there can be different answers. The supervisor will seek to avoid blame, while managers often blame carelessness. Computer based reporting of accidents can produce illuminating analysis.

Studies of organisations and systems focus on the management of safety, and a safety culture. In a booklet "Organising for Safety", key issues included resources, participative relations, visibility of senior management and the balance with the need for production.

A multidisciplinary group at McMaster University developed a conceptual framework, which includes management and labour in the workplace, to which they bring values and priorities. In the workplace there can be socio-technical and industrial relations contacts, and workplace health and safety. The workplace is not in a vacuum, but is affected by regulatory context and economic pressures. Health and safety performance can be explained in terms of this range of factors.

A number of factors in workplace organisation are seen as affecting OHS: organisational structure, philosophy on OHS, labour markets and unions, internal responsibility, demographics, risk conditions and financial performance. Although the relationship was not statistically significant, there was a tendency for more profitable companies to have better safety records. Studies on organisational factors were ignored for many years, and little research was conducted until the late 1980s. The McMaster group looked at the conclusions of previous studies.

Variables consistently related to lower injury rates include training and managerial style and culture, which include empowerment, long-term commitment of the workforce, and good relations between management and workers. Organisational philosophy on OHS was significant, including delegation of safety activities, an active role of top management, regular safety audits, evaluation of occupational safety hazards, monitoring of unsafe worker behaviour, safety training and employee health screening. Post-injury factors include modified work provision. Among workforce characteristics were low turnover and seniority, while other factors include good housekeeping and safety controls on machinery. However, correlation is not the same as causation. The different studies had involved varying degrees of multivariate analysis, and the factors listed can be confounded. Each study used a different questionnaire.

The next issue was to assemble data from interventions, and the available literature, together with the estimated effect size and types of intervention, based on the work of Guastello. He arrived at an average effect size for each type of intervention. Some were interventions aimed at the individual, dealing with personnel selection (insignificant effect), exercise and stress management (modest), and behaviour modification (substantial). Interventions aimed at specific types of accidents cover near-miss reporting (very minor), technological interventions (reasonable) and poster campaigns (modest). There was spirited debate on near-miss reporting, noting that the data came from interventions, and that the starting points are not stated.

Interventions at the group level include quality circles (significant) and autonomous work groups (where in a single study the control group got worse). Interventions at the workplace level include management procedures (strong), housekeeping (strong), the International Safety Rating System (modest), safety committees (modest), comprehensive ergonomics (strong) and cooperative compliance (strong). Interventions at the jurisdictional level vary: it appears that OSHA in the USA had limited effects, with more impact in Finland. It was suggested that the workshop could add to this picture, drawing on wider experience.

Choice of intervention may be affected by the burden of injury (severity and frequency), cost-effectiveness, short-term as against long-term effects, allocation of resources, cultural and contextual acceptability of particular interventions, the availability of proven solutions (which might be at a small scale, as opposed to globalised expert opinion presented as best practice), attempting to secure management commitment (before, rather than after, a tragedy), temporal preferences or fashions, and understanding of "causes". It is a matter of devising, implementing and evaluating interventions. When we consider the application of current knowledge, there is a gap between what we know about prevention and what is being done.

Is it a question of trust or sanctions such as prison? Opinions are divided on executives' and managers' responsibility. Evidence on management principles and safety suggests that there is little scientific basis for auditing practice. Should they have such a basis? There was resistance to trials in the medical (psychiatric) field, noting the limitations of statistics and of experimental approaches by practitioners, but these have now been overcome.

The need for evaluations is highlighted by education on safe lifting, ergonomics and high school driver education programmes, where effects have been limited. Some may even do more harm than good. The literature is fuller of descriptions than of evaluations. Potential usefulness of interventions can be tabulated, to help attack particular problems.

Is this an area where science can provide answers? Is safety really a matter for the workplace practitioner? Has a great deal of money been wasted on inappropriate projects? Have we attributed too much importance to science?

How do we decide whether evaluations have been done well? He considered:

1. Programme objectives, logic and conceptual basis

2. Study design: was an experimental design employed?

3. Programme participants: were they fully described?

4. Intervention description: were the intervention, context and implementation described?

5. Outcome measurement: were outcomes measured by valid methods?

6. Qualitative data collection: were qualitative methods used?

7. Threats to internal validity: was bias addressed?

8. Statistical analysis: were the appropriate analyses used?

9. Interpretation: was the interpretation of outcomes sound?

10. Conclusions: were the limitations addressed?

Issues for discussion included:

- Barriers to adopting interventions

- Barriers to evaluation

- Poor evaluation and no evaluation: which is worse?

- How to educate workplace parties and policy makers

- Randomised trials: when and how

- Evaluating policy and legislation

- Getting in the door of companies making changes: it is hard in North America

- The importance of workers' involvement

- Likelihood of completing evaluation (companies can go bankrupt)

- Vested interests
- Some interventions are multifaceted
- Uniqueness of interventions: generalisability

The choice of intervention is not always straightforward: how do we decide what approach to take on back pain?

There can be alarming evidence of conflict of interests in areas such as pharmaceutical research: a study checked on relations between researchers and the manufacturers concerned. The correlation was close: researchers avoid biting the hands that feed them. Perhaps safety should be seen in a broader context than scientifically evaluated interventions.

Examples of Interventions

Safety in the Finnish Construction and Metalworking Industries

Heikki Laitinen talked about safety in the construction industry in Finland, where there had been a serious recession. There has been economic recovery, but concern over quality and safety. A safety competition was introduced, between construction companies, and involving trades unions, as well as insurance companies. The evaluation is by the Institute of Occupational Health. Safety seminars are organised, with awards for the best companies. Safety inspectors visit without notice, monitoring the work environment, auditing safety plans and gathering accident data. The walkthrough method includes 100 observations and 6 major safety aspects, calculating a safety index. Reliability of inspectors' conclusions is good. The methods have been used since 1993, with over 12,000 observations in 1997, and an improved safety index, including a fall in accident rates.

Kari Häkkinen argued that the figures require further analysis, checking against compensation claims. Paavo Kavisto noted that increased effort, and an intervention such as this competition, can make a major cultural difference and reduce unsafe acts. Tord Kjellström argued that the question is what use is to be made of the figures. Perhaps different figures need to be communicated, highlighting the important lack of safety. There was debate about reporting non-events.

Heikki Laitinen reported on analysis of safety in the metalworking industry, highlighting the work environment and psychosocial factors. Questions concerned influence at work and relations with managers and supervisors. It was concluded that psychosocial factors do not affect safety, but actions after accidents. Tore J. Larsson noted that employers with a better psychosocial climate tend to be safer: it may be a co-variant with work safety. Heikki Laitinen was asked about the control group, and noted that small companies require separate treatment.

The groups were given issues for discussion on evaluation of safety interventions.

Interventions and Learning Organisations in Hospital Safety

Andrew Hale presented a problem of evaluating interventions to prevent injury in hospitals. A great deal of faith is being placed in learning organisations as a way of improving the management and prevention of occupational safety, health and environment. Learning organisations are said to learn from mistakes, and therefore install registration systems to capture problems. The study was concerned with patient safety in hospitals. The evaluation encounters the same sort of issues as the evaluation of audits or other elements of a safety management system. The system encourages reporting, so the number of reported incidents increases, and evaluation can give negative effects. What ways are there of assessing the effectiveness of introducing incident registration and analysis systems?

Liability for patient injuries, which might be one of the explanations of previous low reporting, had not been an issue. There is an overall database, and each week incidents are selected for discussion by the professional team. Evaluation is concerned with changes in practices and attitudes.

Past research experience in delivery wards suggests that different categories of staff see risks differently. There are no standard weightings. Awareness of a risk affects behaviour. Information about risk factors will emerge from increased reporting, but there can be problems of categorising risks, when there are issues of perceptions and attitudes. Quantification would be difficult in the short term, but patient satisfaction was a possible focus. Incident reporting was introduced to build a learning organisation. This work should be seen as action research, with perhaps insufficient measurement before the intervention.

Tore J. Larsson described hospitals as almost military in organisational structure, with clear lines of responsibility and accountability. **Andrew Hale** described the classification of patients in levels of risk, but the operation processes are not themselves analysed in functional terms. One outcome is the realisation that modifications can be made to procedures. A scientific tool for incident reporting is not the same as changing management practice. The origin of the project was a heart surgeon interested in professionalising processes in the hospital. So what should be evaluated? What was the effective start of the project? This is not a laboratory exercise, and the limits of measurement are extreme. Generalisation is difficult: hospitals are all different, as are departments and individual staff.

Communication is important, and there are complications of researching in areas of professional activity. Some of the lessons may be transferable. It is difficult to write about such matters for journals, with the contrast between scientific approaches and professional cases. **Tore J. Larsson** highlighted the complexity of socio-technical systems, and favours writing up cases over several years. This does not have to constitute boasting by safety officers, but can involve qualitative scientific methods. Experienced professionals can identify, from their experience, the key area of risk from which a base of measurements could be built.

Andrew Hale reported on the problem of observation as a non-specialist, but thought there would be scope for time based research. There are methodological problems in studying medical accidents, which raise different issues compared with occupational health. It is the behaviour of the person engaged in work that causes the accidents.

Mette Dyhrberg outlined a proposed anthropological approach to safety, listening to stories collected on field studies, and based on changing the culture. She may have to persuade companies that her activities will improve their safety culture. **Tore J. Larsson** said that the issues were about attitudes and norms. Social psychology provides a wealth of literature, including the work of Erving Goffman.

A Polish Hot Strip Mill

The other group considered the case of a Hot Strip Mill in Poland. There were more accidents during morning shifts, especially in the second hour of work. Researchers studying data over 38 years attributed this to an enforced early start to the day. The night shift had less accidents, possibly due to a slower pace of work, and informal opportunities for sleep. During the cycle of shifts the third day was the most dangerous, and accidents were reduced on Sundays. Peak times of year were April and October, and with more accidents in the summer. Older workers resist change to their shift patterns. There has been no intervention to date. More information might be collected about the activities at the times with peak accident levels. It may be a matter of temperatures, cycles of processes, or work rhythms.

Danish Nursing Homes

Ingrid Christensen sought advice on an evaluation of work on nursing homes in Denmark. In 1996 the Danish Parliament called for improvements in the working environment, including removal of the range of occupational health factors. There is a campaign on the transport industry, and on particular issues such as lifting. Research has shown particular problems for employees in nursing homes with physical or mental handicap, cased by heavy lifting and poor working postures. There are also psychosocial risks, and indoor climate problems. A campaign is planned for 1999, and will involve discussions with the relevant organisations of employers, supervisors and employees, and senior decision makers. Space and design issues will be set out for architects. Universities will be involved in addressing work environment issues. The national programme will involve a range of means of reaching target groups, seeking cooperation of the social partners. New inspection methods will be used, and information materials will be developed. Measurable concrete objectives are being designed, in terms of company performance in areas such as ergonomics. How are these to be measured? How should the performance of the campaign itself be measured?

This is a Danish initiative, intended to advance good practice. Given the finite resources available, it might be wise to operate random trials. **Kirsten Jörgensen** explained that part of the motivation is to raise the profile of nursing homes with the social partners and the wider public. The Working Environment Service will collect data, but an external evaluator has not yet been appointed, thus cannot compare past and future situations.

The groups proceeded to discuss the case.

The programme is an intervention, provoking interventions and change, from which data may be gathered in an enhanced reporting process. Inspections will not cover all nursing homes, but at least "a cocktail" in each municipality. There will be a press and media campaign, which follows a campaign on psychosocial factors. There are different perceptions by staff, and relations of patients. The intended outcome would be a reduction in occupational diseases and accidents, but in the shorter term better identification of problems: it is a matter of safety management. An initial survey was proposed, and the development of matched "cocktails" of nursing homes, though the media campaign would affect everybody. It will be necessary to know what activities have taken place, and then what the long-term effects may be.

Use might be made of existing survey processes, but normal inspection data is likely to be flawed.

Comparative lessons were learned from an evaluation of improvements in Finnish metal companies, introduced by **Heikki Laitinen**. It is useful to measure effects of different parts of a complex campaign. It was agreed that there are dangers of bias if the evaluation is linked to the inspectors undertaking the work. Will the efforts in the intervention be sustainable in terms of building self-reliance? The political dimensions cannot be ignored.

Tore J. Larsson suggested that the workshop network group might be used for ongoing advice, via email. He closed by thanking the workshop organisers and participants. He noted the role of the International Commission on Occupational Health. The keynote paper and other submissions will be published in the Safety Science Monitor. There will be a scientific workshop summary and a popular report.

Workshop Participants

Ingrid Christensen, Danish Work Environment Service, Denmark
Mette Dyhrberg, Technical University of Denmark
Ad van Duijn, Ministry for Social Affairs and Employment, Netherlands
Inga-Lill Engkvist, NIWL, Sweden
Richard Ennals, Kingston University, UK
Tomasz Gdowski, National Labour Inspectorate, Poland
Kari Häkinnen, Industrial Insurance, Finland
Andrew Hale, Delft University of Technology, Netherlands
Kirsten Jörgensen, Danish Work Environment Service, Denmark
Tord Kjellström, University of Auckland, New Zealand
Paavo Kivisto, Assistant Deputy Minister, Ontario Ministry of Labour, Canada
Heikki Laitinen, Finnish Institute of Occupational Health, Finland
Mats Lignell, NIWL, Sweden
Halszka Oginski, Jagiellonisn University, Krakow, Poland
Simo Salminen, Finnish Institute of Occupational Health, Finland
Harry Shannon, Institute for Work and Health, McMaster University, Canada
Eugeniusz Sulkowski, National Labour Inspectorate, Poland
Arne Wennberg, NIWL, Sweden

Reflections on the Workshop

The workshop had more of a tutorial style than previous workshops, with each day led by a keynote speaker and based on activities that were oriented more towards learning than towards policy. The workshop leader and keynote speakers are all based outside Europe, in the English speaking "Old Commonwealth" nations of Canada, Australia and New Zealand, though their origins were European.

The presumption was of a gap between the worlds of science and of decision making, though participants would define the gap in different way. This was a theme of the workshop on "Research Diffusion", but here it was recognised that thought must be given to communication strategies. Political discourse needs to be able to take account of clear scientific results in a way that is linked through to sustainable action. This represents progress beyond debates purely in terms of market forces. This has previously been justified in terms of the need for solid quantifiable data. When such data is provided from the previously "soft" human perspective, it must then find a framework in which to fit. There was repeated debate about what constitute appropriate scientific methods, when seeking to evaluate interventions which might be seen as constituting action research.

Can the social dialogue deal with such material, or must it be left to public health? What practical impact will there be? How could European directives be developed, in light of disparate circumstances and data?

One challenge of the workshop is to provide data for social policy discussion during the Swedish Presidency of the European Union in 2001, which might influence EU decisions. What information is being collected and quantified? If our categories are not the same across Europe, how can we arrive at reasoned and consistent conclusions? There is no universal agreement on what constitutes occupational trauma. There is no standard set of compensation arrangements. Data are not comparable. In most cases decision makers give little attention to epidemiological data. There is not the basis for a level playing field.

Tore J. Larsson's case studies of the farmer, concreter and apprentice were qualitatively effective. They captured characteristics of people in each country, where the cultures and the remedies available are different. One approach to accident prevention would be to take such case studies as the starting points for scripts for radio or forum theatre. The scripts would be different for each culture, but the underlying safety message could be the same.

The contrasting perspectives of the keynote speakers stimulated vigorous discussion; the practical examples brought by participants grounded the arguments in the complexities of the workplace; and the final sessions were spent designing evaluation procedures.

Information Society

1. Research Dissemination

The workshop was led by Elisabeth Lagerlöf and Markku Aaltonen, and held in Brussels, 23–25 November 1998, at the Office of the Swedish Trades Unions. It was organised jointly by the National Institute for Working Life, Stockholm, Sweden; the European Agency for Safety and Health at Work, Bilbao, Spain; and the Nordic Institute for Advanced Training in Occupational Health, Helsinki, Finland.

Monday 23 November

Introduction and Welcome

Elisabeth Lagerlöf and **Markku Aaltonen** introduced the workshop, whose participants are experienced research managers, information specialists and OSH experts, with national and international roles. Research dissemination has been identified as a key issue by the heads of the major research institutes in Europe. The workshop was intended to allow the sharing of experience, the formation of a potential ongoing network, and the recommendation of future strategies.

Elisabeth Lagerlöf set the workshop in the context of the Work Life 2000 programme, which was introduced by **Bengt Knave**, vice-chairman of the organising committee.

Markku Aaltonen
Research in the Service of Occupational Safety

Markku Aaltonen outlined the work of his information project at the European Agency in Bilbao. He was concerned to identify future research needs and priorities, to collect and disseminate research information, to initiate and support Internet-based developments, and to support research programmes concerning work-related upper limb injuries. The approach was cooperative research with an emphasis on practical solutions. The agency works in a decentralised manner, through networks and tripartite focus points.

He discussed the importance of validating research results, in institutional, scientific, technological, pedagogical and media terms. Drawing on a Finnish model that was frequently used during the workshop, he saw the key relationships as being between research units, and users, aided by facilitators and underpinned by financial support.

Elisabeth Lagerlöf
Research and Communication of Research and Research Results

Elisabeth Lagerlöf is concerned about the mismatch between the optimism of senders of research findings and the use that is made of research by the intended audience. This is a general problem, with particular applications in Health and Safety, where the political consequences can be serious. There has been a lack of empirical research on who wants what information, in what form and at what time.

She presented science as a product, as a source of evidence, and as a method. Research could be Commissioned as a means of rationalising complex problems, in the hope of producing better quality (seen in accident prevention and the reduction of injuries), and more generally as a resource for culture and democracy, in which participation is important.

Access to scientific research is increasing, but with uneven distribution of access. Crucially, non-possession of scientific knowledge is not recognised by those who lack it. Access is not in itself knowledge, and knowledge does not necessarily include understanding. To be effective, knowledge needs to include a theoretical framework, social context, and the support of appropriate discipline. Knowledge can take different forms: direct (in my head), indirect (who can I call on for advice?), paradigmatic (ways of motivating indirect knowledge) or fiduciary (who can I trust?). Underpinning knowledge we need structure and imagination, provided through education.

She considered the various uses of research. There has been considerable work on the dissemination of innovation, but with insufficient attention given to the barriers encountered. The same problems apply to work, where research dissemination is seen only as an Information Technology problem, i.e. use will increase with better layout, shorter text etc. The first studies on why scientific research was used started with the negative result of research use in the 1960s. Social policy used a problem-solving model and there was an elevated role for the social sciences in support of decision makers. In reality life was different, and decision makers made little use of research results (indeed, we could ask whether it would be rational for them to use such results).

What, then, is the use of research? It can be used to produce recommendations for action, to enable decisions to be made which reflect research results, to improve future decision making, or to help change the way in which questions are viewed. She quoted the work of Weiss, who identified the uses of research as instrumental, political, pedagogic, interactive or tactical. Communication theory describes the sender, the message, the channel and the receiver: who says what, to whom, how, why, and with what effect. This suggests that the messages from research need to be adjusted in form according to the intended user. Such a model is far too simple to be of practical use to us in considering research dissemination.

Considerable demands can be made on research information: it should be subject driven, easily available and action related (but this is not always possible). Research can be seen as formal, factual and relevant, but not as simple and unproblematic. The response to research will vary with the perspective of the reader. Uncritical scientific papers can descend into scientific incest. There are issues of democracy and access: information, feedback and an input from the general public are required. As an outcome, research should be better integrated into decision making processes.

How, then, do we proceed? The general strategy is to combine traditional information with improved dialogue with the user and accessibility. The outcomes of research must be translated or re-expressed for the general population, specialists and decision makers. Clear presentation is not enough, we also need to explain the significance of particular findings, and to stimulate the reader to pursue the research further, providing appropriate references. In order to arrive at solutions involving research we tend to need to use brokers, who assist users in problem solving. Some help can come from action research, but there are difficulties in disseminating highly specific case studies, and collaboration with consultants can be helpful. Information needs to be targeted, both in message and form, for particular groups and problems.

Elisabeth Lagerlöf concluded by asking how far, and by what models, could a European body or a national research institute disseminate its results.

Discussion

Kevin Gardiner sought to clarify the backgrounds of workshop participants, who were drawn from research institutes, public authorities, trade unions and decision makers, and were not simple end users. **Hannu Stålhammar** pointed out that the same person could play different roles. **Kevin Gardiner** emphasised the advisability of involving users in research projects from the start. **Jean-Claude André** noted that research could lead to action, but action could also lead to research, and there can be clashes between research and ideology. **Marc de Greef** challenged the idea of research as a single concept, and argued that it varies in nature, sometimes available, sometimes secret. We need to classify the types and consider user feedback.

Michael Claessens, DG-XII
Methods and Tools for Dissemination of Research in EU's Framework Programmes on Research and Technology

An account was given of information and communication in the particular context of the Fifth Framework programme, which has recently been approved, continuing the pattern of increased European Union expenditure on research. Themes include Quality of Life, Competitiveness and Small and Medium Sized Enterprises (SMEs). The objective is to place research at the service of the people, addressing problems of society, and in particular dealing with the needs of target groups, such as science, industry and end users.

Information is disseminated through a number of means, including the World Wide Web, databases, innovation relay centres, publications, information centres, agencies within the member states, chambers of commerce and individual emails. In any given year there may be 10,000 projects running, chosen from 24,000 proposals, and with each project involving an average of 5 organisations and 3 member states per project. The Fourth Framework Programme involved some 500–1000 publications and 100–200 workshops per year, involving 10,000 participants. The better known the programmes are, the more likely they are to receive increased funding.

DG-XII issues communications to inform the public, whose general level of awareness is otherwise low, they organise key issues events, develop appropriate dissemination tools, and set the research in a European Union context.

The problem is not a shortage of information, but effective communication between scientists and decision makers. Priorities include the harmonisation of initiatives across the European Union, the encouragement of results, the involvement of the public, the exploitation of dissemination tools, and the improvement of internal communications. At its best, the European Union should be able to out-perform the United States and Japan. There can be resistance from scientists regarding communication, and there is pressure to publish results through journals such as *Nature* prior to other disclosure. On occasion scientists will be deliberately confused and confusing when talking to the press!

Discussion

Marc de Greef declared that he is fed up with success stories, as published in EU public relations material. Life is not always successful, and we should not be ashamed to describe difficulties. **Michael Claessens** said that the intention is to give a positive account of research, and that every project has a technology stimulation plan. The objective is to enhance technology transfer. **Kevin Gardiner** noted that projects are stopped if they do not meet their objectives. Work has been done to facilitate communication: he gave the example of the 21 Project, with environmental information for small firms, where ways had been found to circumvent some of the language problems and facilitate effective disclosure. He reported that a new Directive on Intellectual Property Rights is under discussion, which should result in more consistent arrangements. Several speakers noted the importance of the CORDIS programme, although those without Internet access at their desks, such as **Gillian Lowe**, derived less benefit. It is hoped that the Bilbao agency will help Health and Safety research results to reach SMEs, who then need to be able to use the information.

Jean-Claude André, Research Director, INRS, France
Transfer of Occupational Health Research to Society: the French Exception

Jean-Claude André spoke from the perspective of research laboratories, where power can be seen in terms of a credibility cycle, which involves securing funding and reputation but may not include contacts with users. Such laboratory cultures are hard to control in a way that brings greater efficiencies for society. Industry takes a global approach, concerned with demand and technology linked to social value, deals with complexity and likes to control relations with research. Researchers tend to adopt a more local approach, with an emphasis on peer evaluation, specificity and academic networking. These two approaches are different and may come into conflict.

In order to advance social change, a top-down approach is needed. We know that change is coming, with clear indications of demographic and occupational shifts, including a move from manufacturing to services. Outside the research laboratory we see new patterns of work organisation, including subsidiaries, sub-contracting, Just In Time, globalisation and the development of new forms of organisation. There is a rapid expansion of part-time work, temporary work, fixed term contracts, early retirement, flexibility, 35 hour weeks, teleworking facilitated by IT, and job sharing.

We have an unstable social system. Risk management is being considered in the context of a shift in social paradigms, a lack of social confidence, with worries over liability, prevention, objectivity and globalisation, all adding to anxiety factors. New scientific findings can add to stress. As the population balance shifts, older in age and with a smaller proportion employed, the pressures of globalisation can lead to exclusion and marginalisation. This amounts to a paradigm shift. The classical approach to risk management involves taking rational scientific action in order to improve the situation, but an atmosphere of collective doubt makes this harder, exacerbated by the media. Instead of confidence in solutions derived from engineers, scientists and statisticians, we face uncertainty.

As we consider research approaches to risk management we envisage less bottom-up proposals, greater use of quality loops and clearer publication of results. Medium term research plans need to include observation of external problems, understanding and moves to improve the situation. This involves a shift of responsibility towards considering the needs of the user. In general this requires the development of partnership, emphasising prevention, basic organisational issues and links with government and industry. Given all this, INRS will continue to be of some help.

Research results need to be more action oriented. Dissemination can take a variety of forms, including good scientific papers for a specialist audience, software for industry, patents, prototypes, and involvement in standardisation and certification. We must expect the business of research to become more complex. We are moving from regulation to advice, adopting inter-disciplinary approaches, developing and maintaining links with the outside world, and engaging in foresight exercises.

Research has to take account of the clients, the users of the products of research. INRS is to some extent its own client, but it also deals with the Caisse Nationale, Caisses Régionales, industry, enterprises and doctors serving 15 million employees, and 6000 occupational health physicians. Given that 70% of French employees work in 1.5 million SMEs, services and publications need to reflect this. There is a range of regular publications targeted at different audiences.

As for the future, the challenge for research is to build up external relations with society, and to develop public confidence. This means publishing credible valid results, which may be easier for chemistry than for stress or ageing. It raises issues of scientific ethics, and requires a process of synthesis for debate and dissemination. Research has to prepare society for the shocks that are to come. This must be done by convincing argument, not by force, and involves a new separation of powers between science and society. Society has to choose. It is a matter of improving dialogue with society at large, and developing new cooperation at a national level and across Europe. Vectors in that dialogue will include journals, the media (for explanations) and Information and Communications Technologies. The whole business of scientific and technical communication is undergoing radical change. We face a large task.

Discussion

Jan Nielsen asked about evaluations of the effect of scientific information on society. **Jean-Claude André** noted that French researchers are civil servants, who find change difficult. Their main dealings are with their paymasters, rather than with users. **Kristina Kemmlert** described the approach at the Swedish National Board of Occupational Safety and Health, where there are three year goals to be met, and surveys are conducted to assess the level of success. **Jean-Claude André** noted the time-scales involved in anticipating the research necessary to produce timely results. The INRS Board includes representation from trades unions and employers. **Elisabeth Lagerlöf** asked if research institutes should tell decision makers what they need to know.

Kevin Gardiner noted that the countries of Europe are all different, for example in their organisation of occupational health. Scientists tend to publish in order to secure funding and peer acceptance, and not to solve external problems. Boards of scientific institutions tend to resist new ideas. It is worth asking how much research gets through to the field. The British Library Association is worried that, in light of the dubious quality of much material on the Internet, scientific communities are likely to publish their material on closed Intranets, restricting access. **Jean-Claude André** argued that researchers need security, and should not be obliged to publish or perish. Decisions regarding publications are made by the INRS Board, based on scientific and political evaluation.

Marc de Greef felt that the discussion had muddled several topics. He highlighted the dominant role given to science in the area of Health and Safety, which is located in a social context. It is seen as technical, managed by institutions, and with problems to be solved by experts. He argued that there is no one best solution, and that we need to be able to move between roles. We need to get closer to users and, as we work at the European level, avoid making all the same mistakes.

Group Work

Elisabeth Lagerlöf introduced the group work, in which the participants formed four groups, considering accidents, chemicals, musculoskeletal disorders and stress. In each case the group was set the same set of questions in the context of sharing experiences:

1.   What different approaches are there?

2.   Who has been the target group?

3.   What was the message?

4.   What tools (media) were used?

5.   What were the results?

6.   What could have been improved?

This was followed by consideration of objectives for future research communication, which should also be measurable.

Accidents

The group lacked practical experience of dissemination, but noted that the Belgians do little, the British HSE publish only some of their work, and the Swedes are full of ideas.

They discussed high risk scenarios such as nuclear war, Seveso and transport disasters. They explored different types of information and dissemination methods. Different approaches include technical solutions, methods of risk analysis, evaluation of societal approaches, problem mapping and statistical analysis, and development of solutions to define a problem. Target groups tend to be researchers and the research community at large; authorities including inspectorates; and insurance bodies.

Accounts were given of the DG-XII SHARE project, with 20 participating countries including Sweden. It was noted that research tends to have a long lead time, while action needs to be fast. All too often Nordic research results are little used. The recommended strategies tend to be of little use to SMEs, deriving from experience of large companies. The British HSE have a tripartite system of committees, with additional involvement of consumers. The discussion closed with a consideration of data, information, knowledge, competence and skill.

Chemicals

The knowledge exchange was based on personal experience. Four approaches were identified:

Databases

These are targeted at scientists, politicians, employers and employees. The media used are paper, CD-ROM and the Internet. The result is information transfer to large numbers of people, and wide use of the information. Improvements would be better information, enhanced linkages between databases, user-oriented design, the development of meta-databases, and the provision of training in how to use the new media.

"Valorisation" Documents (Adding Value)

These documents are intended to provide recommendations and guidelines, together with information. The targets are authorities, experts, employers and workers organisations. The results support the change process and improved prevention and control. The same media are used. Key improvements would be in timing and prioritisation. An example was given of a new substance used as a tunnel sealant which was well known to be neuro-toxic; but this knowledge was not available to the new user group.

"Vulgarisation" Documents (Popularisation)

This was the term used (the English prefer "popularisation") to describe the adaptation of scientific content and form to meet the needs of target groups. This involves a number of activities in parallel including moving, where possible, from technical text to pictures. For many institutions it is a problem persuading scientists to write non-scientific papers, for which they may gain less professional credit.

Legal Documents

Legal documents are another means for dissemination of research results. Amid all the arguments on deregulation, it was agreed that laws can help underpin arrangements.

Other Issues

Especially in Europe, one might find it not very difficult to do the same work again and again in the different countries. So it would be helpful to have e.g. in a "meta-database" the common facts available. But one must accept that the different countries have their different conditions. Thus the "meta-database" has to be adapted to the specific situation in the different countries. Even different perceptions of risk have to be considered.

Musculoskeletal Disorders

As if in response to the previous group, the report of the musculoskeletal group was based on two diagrams. The first showed how musculoskeletal research findings could be processed into the form of regulations and then simpler guides, deriving benefit from analysis of practical workplace situations and industrial experience. Ideally users, manufacturers, distributors and purchasers would work together to contribute their feedback to the research process, while the findings also stimulate pedagogic and motivating discourse. The second diagram was that used by **Markku Aaltonen**, amplified by an account of the Finnish Work Ability programme. Among the outcomes of this large-scale long-term programme are new measurements, providing the means of predicting work ability for particular occupational and age groups.

Stress

The stress group identified individual and collective approaches. Stress is a widely used term, and definition is difficult, meaning that measurement could also be less than straightforward in its implications. Changes in work organisation can have adverse consequences in terms of occupational health and safety. Australian research has highlighted problems for contingent workers, who also tend to lack trade union representation. Depending on the occupational context, stress can be seen as positive or negative. Recent public examples, such as of the Norwegian Prime Minister, are making stress more acceptable, and bringing home the point that it is not a minority problem, something experienced by failures.

The dissemination model put forward by **Elisabeth Lagerlöf** was seen as too linear, and the group preferred the Finnish model, offered by **Markku Aaltonen**, noting that the details of institutions and targets would vary greatly between countries. Overall the favoured approach was networking and cooperation: collective coping with increasing levels of stress at an organisational and societal level. Stress is not restricted to those currently in employment, and there was consideration of problems of the unemployed. We need to identify measurable indicators, and to consider stress within an overall context of health promotion.

Tuesday 24 November

The focus of the second day was on action.

Paul Schulte
Dissemination, Receipt, Utilisation and Impact of Information

Paul Schulte spoke from experience at NIOSH. There are inherent tensions, as seen in the NIOSH vision statement: "Delivering on the Nation's Promise: Safety and Health at Work for All People, through Research and Prevention". Research and prevention, or dissemination, are in tension. Some think that information thrown from an ivory tower is disseminated. However, it must be an active process. He outlined the range of publications, covering peer reviewed research, translation and policy documents. Direct dissemination is via journals, hazard evaluation reports, patents, research partnerships and databases. There is a need to know how effective the dissemination is, so bibliometrics and citation analysis are used. This is not enough. Bibliographic mapping enables one to discern patterns of influence.

The vulgarisation, or bowdlerisation, process includes telephone hotlines, the Internet, alerts, brochures, worker notification, curricula for workers, employers and OSH specialists, and exhibits. The Internet site will be developed with contractors, but there have been millions of hits on the site, doubling each year, with 7 million in 1998. Every document published is also put up on the Internet. Only 15% of the US population uses the Internet, with lower use among blue collar workers. It is used by the mediators.

Valorisation is through criteria documents, hazard reviews, current intelligence bulletins, best practice conferences, curricula development, skills standards, policy development and technology transfer. It is hard to get scientists to work on vulgarisation and valorisation materials as part of their priority work, and a level of skill is required. We need to consider what works and what does not work. NIOSH are working with vocational and technical education, and with skill standards initiatives. Health and Safety needs to be at the core of future models of work.

The old way assumed one way, with a broad reach and largely use of paper. Now there is social marketing, based on audience segmentation, in electronic media. The work is interactive, "Just In Time". There is a new focus on small business, with material that is easy to read, targeting decision makers at the time they need the key information. The worker needs to be an active participant. Many workers in the USA are illiterate, semi-literate, or foreign born with other first languages. The audience needs to include insurers and those who finance projects.

He presented a cyclical model linking research, publication, adoption of findings, action/regulation or training, reduction of morbidity and mortality, surveillance (of diseases and hazards), and stakeholder inputs. The research is not merely academic, but must move people to action that makes a measurable difference in an ongoing process. The stages are not independent, but linked each to the others. Research enables the setting of priorities, given finite resources. The model occurs at multiple levels: societal, organisational and individual. We need a new kind of research, following basic and intervention research. We need to deal with enabling factors. Given that we know what needs to be changed, why have these changes not taken place: what are the obstacles? What are the economic and political dimensions?

He distinguished kinds of research output, in temporal sequence: immediate, intermediate, pre-ultimate and ultimate. This can lead to an integrated figure of merit as an overall assessment of merit, integrating downstream stages, incorporating quantitative and judgement elements, research accomplishment and cost. He then looked at levels of intervention research (policy, programme and work site) and three categories (development, implementation and evaluation). Resources need to be put into dissemination, for the reduction of morbidity and mortality, and with indicators to monitor progress.

Discussion

Elisabeth Lagerlöf asked why researchers should be writing for the next stage of audience. Some are good at this additional task, and some are useless. There is no training or education in this area, as it has not been seen as important. We underestimate the skills required. Perhaps we need to educate particular people for this task. **Paul Schulte** agreed that this had been a problem. With enough resources, people could be hired for vulgarisation. Valorisation requires scientific interaction, involving basic science. There is no longer the luxury of being a basic scientist devoid of context. There must be interaction with others. **Elisabeth Lagerlöf** noted the imbalance of resource allocations. **Paul Schulte** agreed that the structure is controlled by the research. It reflects "who we are and what we do". There are different world views. The balance of opinions is shifting. **Lars Grönkvist** asked whether there are advance criteria concerning what research is to be continued. **Paul Schulte** indicated that there has been progress: a topic concept memo precedes the vulgarisation and valorisation.

Kari Heftye Skollerud asked about education for health professionals. **Paul Schulte** indicated that more needs to be done.

Markku Aaltonen asked about the future. What about the importance of the Internet?

Paul Schulte sees less reading of papers, but a need for more assessment of users of the Internet, leading to tailoring of strategies. There is a danger of overloading people. Training is required, more than people tend to think.

Gisela Kiesau asked about Health and Safety inputs into vocational education. **Paul Schulte** described the early stages of work on standards and skills. **Gisela Kiesau** argued in favour of integration into existing curricula. This causes tensions with those in charge of the curriculum. **Paul Schulte** described work with high schools, using a science base and accompanied by evaluation.

Lars Grönkvist
Popularisation of Research – the Media and the Press

Lars Grönkvist had a background in journalism and research before joining NUTEK. He is concerned with media as the main source for ordinary people to follow research and development. Few people derive the information from education or experience, but it is usually filtered by journalists. Research and development are not often covered in the news, and there are few science journalists with experience of research and scientific methods. In Sweden only the two biggest papers have science reporters, with sections in the entertainment section on Sundays. Coverage in the media is rare: the potential target group is seen as small: there is no profit to be made. In the 1990s coverage has improved, with focus on biotechnology and space, dealing with perceived risk. Global warming and other environmental problems are gaining attention, but occupational health and safety has had low priority, not mentioned in surveys of priority topics.

The meeting between science and journalism can be dramatic, after long distrust, but they need each other. Science needs funding, and journalism needs expert comment. Lack of knowledge about press coverage and science practice is a partial explanation: neither understands the work of the other. Journalists start with conspiracy theories, and need education and training to broaden their perspectives. Scientists never know how journalists will use information, but it will be like that of other professionals, developing and supporting arguments. There is a culture clash: researchers prefer abstract information, with principles, theory, a professional audience, detail, plenty of time, and caution. Journalists favour a concrete approach, with events, an applicable account, and a wider audience. News criteria are clear, with politics, economics and crime at the top, with local interest, sensation, visible people involvement, negative features and prominent sources. Stories tend to arise from internal leaks and tensions in organisations.

Three examples were given:

- Exposure to electro-magnetic fields leading to cancer (where US studies support Swedish findings) where the issue was of promotion rather than initiation, effects in combination with other factors. The story was eclipsed by a major political story which broke overnight. Soon afterwards a junk news story suggested "400,000 at risk of cancer" simply based on the number living near power lines. The important point was that a number of people were involved.

- Hypersensitivity to electro-magnetic fields is an increasing problem, but there had been no established scientific evidence. Self-reporting was the basis. Scientists were seen as bad guys, and lobbyists as good guys, with effective demonstrations. Suffering people were exploited for entertainment. The reports led to more self-reports. Researchers were forced into further action, and the subject disappeared from the news. **Bengt Knave** announced that the subject has returned with coverage of the impact of mobile phones. The focus of interest has shifted.

- A third case concerned coloured diesel fuel, to stop tax evasion. Stories about allergies spread. "They are not allergic to diesel, they are allergic to taxes" was the conclusion of the scientists.

It is still a challenge to work as a press officer in science and technology. The media can force issues onto the agenda. They can provide information, but are not good at changing attitudes. To be effective with journalists, be prepared. Choose messages. Tell a story. Do not comment on the work of others. Do not lie: journalists like to expose liars.

Discussion

Bengt Knave described experience of having qualified journalists as part of the team at the Swedish National Institute for Working Life, pioneered by **Elisabeth Lagerlöf** and **Lars Grönkvist**. **Paul Schulte** reported that NIOSH have recruited a journalist to head public relations. **Bengt Knave** noted that journalists could make good use of interviews. **Hannu Stålhammar** agreed that the press have little interest in occupational health and safety, but described his experience when a paper published a version of his model without his knowledge. It can be a matter of collecting the right people, and knowing who your material is being sent to. **Lena Sperling** asked whether researchers have a right to read what has been written about their work. There can be distortion. **Lars Grönkvist** indicated that you can ask to check, but the article is the responsibility of the journalist. **Gillian Lowe** described the work of the government information service in support of HSE. Professional assistance is available, and training is provided, which is seen as important. Even senior people can have problems with aggressive British tabloid newspapers. **Joachim Lambert** reported that in the past his organisation were not concerned with the press, but now there are regular briefings. This is cheaper, and more effective, than employing a large staff. **Kari Heftye Stollerud** was a journalist, and described strategies for placing stories in the right places to reach the target audiences. **Paul Schulte** saw the need for more public relations work. **Lars Grönkvist** argued against press conferences and press releases: it is better to use contacts with known journalists. It helps to write the press release for the target audience.

Gisela Kiesau
Education and Training as a Tool for Research Dissemination

Gisela Kiesau is responsible for research applications in companies of work in occupational health and safety. There is indeed a tension between researchers and "dissimulators", particularly in language. She has a background in economics and social science, with experience of research. She described work on the humanisation of working life, where money was spent but practice appeared little affected. A new application department was created, and she has eighteen years of experience of success, tension and political problems. It is a question of bargaining. The Federal Institute of Occupational Safety and Health is directly linked to the Federal Ministry of Labour and Social Affairs. The Institute develops solutions by applying research results from the work sciences. For research and implementation, there are two other budgets concerned with dissemination, separate from research budgets. Today there is less money, but more clients. This means finding new strategies.

Since 1980 the role of the Institute, set down in law, has been to support the Ministry in fulfilling Health and Safety legislation through research and dissemination. There are three instruments: collection of codes of practice and publication of quarto brochures in series on health protection, technology, qualifications and organisation. Research results have to be published as guidelines or manuals, and then seminar concepts are planned, for delivery by others (3000 participants per year). A yearly catalogue is issued, setting out the programme and prices. Practical experience of seminars is vital if concepts are to be developed. There is a new seminar concept for Occupational Health and Safety engineers, twice as long as before.

Codes of practice can be developed earlier than fully normalised systems. Abbreviated summaries of results of research are less than 20 pages in length, and the researchers are asked to undertake the work, with subsequent editing through dialogue. Looseleaf publications are sent to 3000 clients. These codes of practice precede regulation, but have a useful role in convincing people of the importance of new research results.

Seminar concepts are derived from research in particular fields. Our first strategy was to offer solutions: but what was offered was not acceptable. The next strategy was to listen to demands. The third approach is a dialogue strategy, starting with the target groups to identify needs and available results. The concepts are then developed iteratively over time, but it is important to discuss and gain consensus. Mediators are involved, with means of dissemination.

The seminars involve three kinds of papers. There are modular sets of papers for participants. There is a pack for the trainer, including media to supplement paper. Concepts are tested in the Institute with the target groups, and returned to the experts after testing. The seminar is then offered cooperatively, enabling delivery in third party factories and the education system. The cooperation partner must provide the rooms and organise the event, and arrange for participants. Funding is joint, based on joint motivation. The seminar concepts are made in a modular way, allowing third parties to adopt them in whole or in part. This is subject to contracts, which cover reports on delivery to target groups and recommendations of future changes: it is an excellent form of dialogue and feedback. Numerous objectives have been covered: EU machinery, VDUs, chemical protective clothing, workplace design for the disabled, chemical exposure etc. After a period the seminar materials are updated. Each seminar concept costs about DM200,000. Development is undertaken by collaborative teams, including Institute staff. There can be arguments, as scientists may argue that they have the ability to do the work themselves; however, the researchers approve the material before it is published.

Discussion

Gisela Kiesau sees competition as healthy. Her courses do not run at a market price, but are underpinned by government funds. Objectives and target groups are different. On occasion there are joint ventures. Jan Nielsen thought that dialogue was being used as an alternative to market mechanisms. Gisela Kiesau gave insights into ways in which the dialogue could be initiated, often through workshops where proposals could be explored, and the design of materials collaboratively agreed. As a result, the publications can be joint. Paul Schulte agreed that partnership with affected groups is vital.

Lena Sperling
From Research to Practice – the Hand Tool Project

Lena Sperling had brought hardware in support of her presentation. She works for Lindholmen Development, owned by the City of Gothenburg, and working in association with the Swedish National Institute for Occupational Health and Lund University. Support had also come from major companies such as Saab, ABB and Volvo.

The research was based on a holistic perspective, including work design, manufacturing processes and work organisation. The tacit knowledge of the user of hand tools was vital: manufacturers had a mistaken view of how plate shears are used. New tools and methods were developed, including new interview methods. Figures need to be given to qualitative values. The award-winning cube model presented the combination of demands of new situations, in terms of force, time and precision.

An applied project with an industrial perspective arose within the Uddevalla factory of Volvo, embodying good ergonomics. New hand tools were developed, but with a small market. The needs of women were taken into account. Typically manufacturers are remote from users. She illustrated a traditional tool for cutting flax, made in close association with users. User requirements should be central: the technical methods of developing good tools are now available. Typically tools in industry are selected by men, but are not always appropriate for women.

The aim of the project was to develop and make available ten new tools. In addition, there needed to be improved understanding of the importance of better hand tools. Actors involved were users, manufacturers and researchers, together with distributors. The groups worked together: selecting the ten tools, developing prototypes, and then introducing the products to the market. The start was with a large scale seminar, where problem criteria were addressed. The Swedish model of participatory ergonomics was followed. Problems were then mapped in different industries, led by users. Many problems were solved by identifying good tools that were already available. The worst were identified, such as the industrial hammer. Women were using such hammers in Saab, and holding them close to the head. Analysis of the ergonomics of hammers showed their weaknesses in terms of balance. User requirements were collected, and press releases were issued, using illustrations of problem tools. Press coverage was good. A good tool check list was used, adapted from Volvo. There were negotiations with manufacturing companies, some who preferred not to be funded, and the full battery of technical measurements was used. Professional users were interviewed, addressing issues of functional quality and professional pride.

Two successful products were presented in diagrams: knives and band cutters. The problem knives have been withdrawn, and the new ones sell well. Problem and prototype tools were compared. The results could be presented in medical terms as required. Scientific diagrams command respect.

The eleventh product of the project was an ergonomic quality declaration. This was resisted by Swedish manufacturers, on commercial grounds. A quality label was proposed.

A training kit for hand ergonomics was a further outcome, with models of collaborative education for consensus building. It is not clear how much the kit is now used in enterprises, but it forms part of current work in universities in the area.

The new industrial hammer (T-BLOCK) was developed according to user requirements. It incorporates counter-balance for easier use, and a quality of design that gives pride to the user. The "erguments" are in terms of health, quality, productivity, total economy and pride. The economics of good and bad hand tools were explained. Advertising material emphasised economic and ergonomic factors. Handles come in three sizes (the smallest called "medium", as men will not buy a small hammer!). Further work developed power tools for healthy hands. **Lars Grönkvist** published a report on the project, and it was documented by a doctoral student at Chalmers University. Information and dissemination was an integral part of the project from the start.

Discussion

Markku Aaltonen welcomed the practical results: hardware is a form of valorisation. He asked about comparative price levels. **Lena Sperling** noted that these are industrial tools, not sold through domestic consumer routes. The problem can be that attractive tools are stolen from the workforce. The answer may be personal tools.

Kari Heftye Skollerud asked about feedback from the users about their tacit knowledge. **Lena Sperling** noted that there was a problem that not all tools could be improved during the project. It is important to satisfy users if they are to be involved in future.

Joachim Lambert asked about the spread of the specification beyond the initial partners. **Lena Sperling** indicated that all Swedish companies were able to take part, and a number of foreign companies were involved in the project. **Markku Aaltonen** noted that related work has been conducted in Finland. **Lena Sperling** said that the collaboration with Finnish companies had been good, and continues. **Gisela Kiesau** was delighted with the report of the project, and described demand for new projects to develop tools in Germany. She has the necessary funds, and does not want to reinvent wheels!

Group Work: Designing Information Strategies

Stress at Work

Objectives

Cultures vary between countries. Messages need to be focused on attention, knowledge, attitudes and actions. In general stress is increasing, and this includes family life as well as work. Stress can be hard to link with indicators, but self-reported stress is useful as a predictor of future sickness. There can be links between stress and cancer, heart problems etc. We can look at stressors: tight deadlines, inappropriate work organisation and work habits, such as working late. People have to re-learn about stress: it can be seen as positive but this can be misleading. Action can be taken at European and national level. Good practice can be helpful, and can mean savings on resources.

Target Groups

Social partners at all levels. We should also cover professional groups, including doctors and teachers, as well as occupational health. We cannot use regulation and control, but advice. Systems need to combine their efforts to meet needs in a coalition approach in dialogue with companies.

Messages

We need a global perspective on work ability, health and well-being. Unemployment can be a special factor. Economic and productivity effects need to be clarified: a stress worker costs the employer money. We need to communicate information about good and bad practice, and how to reduce stress. Conflict is not necessarily a problem, and the emphasis needs to be on handling conflict. We do not lack information; the problem is handling the information and giving it to the right people at the right time.

Methods and Tools

The group favoured a dialogue with users. Training must involve users in the development process. IT-based reporting systems could be useful, taking advantage of the Internet. We are concerned with developing tools for learning organisations. This could involve Forum Theatre and other forms of the arts. Work health promotion has an important role, as part of a recent EU policy statement. There was debate about general campaigns and their effects: this does not seem to be a sensible focus. Effort should be on the human side, in the context of learning organisations. "Vulgarisation" is perhaps a European joke. "Popularisation" is not quite the same. It is a matter of translation between cultures.

Musculoskeletal

The objectives were early detection, early action, communication and improved safety. This means continuous dialogue and early diagnosis. Target groups at company level are workers and employees. At national level there are decision makers. There is also the international level. At each level we need to address builders and architects.

We want a common vision as a basis for research. There needs to be a company culture, identifying problems inside and outside the workplace. We need to consider new technologies and approaches to training. Preparation must begin in schools, working together on occupational health and safety. Prevention is profitable. We need a holistic, integrated, multidisciplinary system. There need to be research findings, political debate and collection of best practice.

Tools and methods need to be developed with OHS and insurance companies. We need to consider the future workplace of 2020, addressing problems of the elderly. This means working with other groups such as physiotherapists, psychologists and social workers.

Chemicals

There is a lot of information available in databases and journals, but there are chemicals which are not well known. They are new and replacement chemicals. The objective is to reduce morbidity and mortality from those chemicals. In addition, to enable the implementation of information in SMEs. How, and to whom should the information be given? The starting point is with trade associations, trade unions and chemical manufacturers. Decision-makers are important (owners of SMEs). The message is in the form of questions. How can we control exposure? How do we know about exposure? How do we help target groups with guidelines? There are questions about saving costs.

Methods should be databases in different countries, which contain very different information, which could provide an overall holistic view. We should share solutions from different countries. We should work more collaboratively, especially with SMEs.

Videos can be an effective dissemination device, together with interactive Web sites built be experts collaborating across countries. Information packages should go to product suppliers. Work needs to be done on "valorisation" (adding value, or validation) and "vulgarisation" (popularisation and translation).

Accidents

The group looked towards the future in a changing society, with an emphasis on changing public services, and on SMEs. There is no evidence of more accidents in SMEs. It was felt that there could be no one overall solution or approach. Target groups are public administrators, public services purchases, authorities and inspectorates, workers, consumers and service providers. The message to public administrators is that safety is a priority: it's your responsibility; poor safety costs you money; contract compliance is an answer.

The method would be through information and media activity, targeting umbrella organisations. Tools are good examples, follow up on examples, and health and safety integrated with quality assurance. The goal was to use science and proven experience. There are poor solutions on offer, from consultants who do not use science. There is a role for national authorities. The whole process is clearly complex, and requires a great deal of thought.

Discussion

There had been a lack of discussion of evaluation, feedback loops and measurable objectives. Statistics are needed at all levels, and instruments are needed to evaluate the material. This coordination and analysis is a task for the European Agency. Some measurable data can be provided, even if measured in different ways, and on a smaller less dramatic scale. There is a need to share approaches to evaluation, with more interchange and dialogue. Information dissemination is an intervention, and should be followed by research. Awareness of particular issues can be measured.

Elisabeth Lagerlöf set some homework. We need to build the basis for a network for future research, looking at issues of evaluation, research use, and time lags. **Gisela Kiesau** reported that in Germany there is a lack of application research. There is enthusiasm for networks, but we do not know what the outcomes are, or how they should be managed over time. How can we reach target groups? We want to be successful, and in our funding proposals we stress success, but we also need to know about obstacles to dissemination, including via networks. There are common problems and approaches. We need to look at problems in application research, and we need to have appropriate evaluation methods in place. There is a case for evaluating media use of research material, tracking advertisements, radio and television.

Gillian Lowe described the evaluation of HSE work in the UK, tailored to local needs, but with good ideas involved. Each year a number of topics are evaluated. Evaluation also provides opportunities for further development. The evaluation is in terms of value for money, use to the community, and effectiveness of campaigns. The work is done by consultants. Measuring societal impact is the hardest aspect. Evaluation is assigned about 2% of the research budget. International peer review is undertaken, but is not cheap. **Paul Schulte** praised protocols and peer reviewed research. This involves resources. The best way of making the case is by dong good work and publishing it.

Markku Aaltonen secured agreement on the homework assignment. **Elisabeth Lagerlöf** welcomes examples and proposals, within fourteen days.

Bengt Knave concluded that the workshop was successful. He had been convinced that the potential was there for an input to Work Life 2000. He could see a new future, as the workshop is the start of a new process. He thanked the rapporteur! **Elisabeth Lagerlöf** and **Markku Aaltonen** had been successful leaders.

Workshop Participants

Markku Aaltonen, European Agency for Safety and Health at Work, Bilbao
Jean-Claude André, INRS, France
Veronique De Broek, Belgium
Michel Claessens, DG-XII
Richard Ennals, Kingston University, UK
Kevin Gardiner, DG-III
Mark de Greef, Belgium
Lars Grönkvist, NUTEK, Sweden
Lars Harms Ringdahl, Sweden
Kristina Kemmlert, Sweden
Gisela Kiesau, Federal Institute of Occupational Safety and Health, Germany
Bengt Knave, NIWL, Sweden
Irja Laamanen, Finland
Elisabeth Lagerlöf, NIVA, Finland
Joachim Lambert, Germany
Gillian Lowe, Health and Safety Executive, UK
Jean Muller, INRS, France
Jan Nielsen, Denmark
Richard Nobbs, DG-V
Dimitrios Politis, European Foundation, Dublin
Milles Raekelboom, Belgium
Marc Sapir, Belgium
Catherine Schlombach, Germany
Paul Schulte, NIOSH, USA
Kari Heftye Skollerud, Norway
Lena Sperling, Swedish National Institute for Occupational Health, Sweden
Hannu Stålhammar, Finland
Marta Zimmerman, Spain

Reflections on the Workshop

If we call the process of translation from scientific to general discourse, "vulgarisa-tion", what is the term for the reverse process? Is it "scientification" or "mystifica-tion"? Are we saying that there are, and must be, two separate cultures? This has serious implications for education, business and politics. It has been a commonplace in the United Kingdom that two cultures have developed, in which more status is ascribed to the arts, business and finance, and it is thought acceptable for educated people to know little mathematics or science. In France, however, there has been greater respect for science, and career structures have been maintained for researchers.

The discussion suggested that the major problems were:

* *Motivating scientists to talk to people outside their closed professional world.*

* *Motivating decision makers to listen to scientists and read their research findings.*

Changing work organisation involves scientists in their capacity as "knowledge workers". They cannot pretend to be detached and objective about a process in which they and their science are inextricably involved. This raises issues about language, explanation and the philosophy of science.

The Scandinavian and Northern European approach might have been thought to have bridged the gap. By having consistent funding for research on occupational health and safety, surely the research results are reaching those most directly affected. Perhaps unsurprisingly, the gap has simply been relocated, so that there are complaints of a lack of communication between researchers and practitioners in the workplace. The results of a given research project may have applications across Europe, and may contribute to diverse dialogue. One general answer seems to be to learn from differences, and to add an international dimension to the dissemination process.

3

Conclusions From the First 12 Workshops

Work

Work is of central importance in the European Union today, and is not simply a matter for market forces. All European Member States are committed to common policy guidelines agreed at successive summit conferences. Working life is more than just the alternative to unemployment, and problems with work have impacts on families, society, the wider economy and politics. Dialogue at the workshops has centred on working life, and how it can be improved. This has involved a range of professionals and researchers, together with representatives of the social partners: employers and trade unions. Each European Union Member State has different experiences to report, although many of the same multinational corporations are involved, and all are covered by the same European Directives.

Workshop Themes

The workshops have been almost free of both politicians and "management gurus", and there has been mature scepticism of over-easy prescriptions. Much of the debate has concerned the pragmatics of combining responses to diverse external pressures, for example on quality, environmental management, and occupational health and safety. There is limited reliance on systems, whether expressed in terms of technology or management. There is pressure to systematise in many areas, but external prescriptions can be inconsistent. Standards and certification meet the needs of regulators, but may have mixed consequences for the organisations concerned.

To date the workshops have not included passionate advocacy of a strong and united trade union movement, but trade unionists have made active contributions to a dialogue where their views are given weight. There has been no presentation of management and business education as a potential major contributor to the improvement of European working life, which perhaps suggests that practical experience of working life should precede management, and that the whole role of management faces major change in a workplace of teamworking and flattened hierarchies. Although numerous officials from European Commission Directorates-General have attended the workshops, there has been no explicit advocacy of strengthened European Directives for the area of working life. It is recognised that top-down imposition is problematic, and that new Directives need to be arrived at through dialogue. De facto the workshops constitute an ongoing European Social Dialogue, in which the emphasis has been on learning from differences.

Some contributions have reflected traditional views of work, based on the exercise of skill in manufacturing. This had been the context for the development of national institutes concerned with working life, while the balance of employment has moved from manufacturing into services, and from permanent full-time jobs into more varied arrangements. Work organisation within and between enterprises is undergoing transformation, and most workshops included some discussion of networking and coalitions. The dialogue has reflected the pace and diversity of change: globalisation, technological change, the increasing coherence of the European Union, and imminent expansion of membership.

Each workshop brings together a new European network, which can use new contacts to build lasting research collaborations for practical benefit. This can take the form of partnerships which are eligible for European Commission funding, or conversations which continue through electronic means. Active dialogue raises possibilities, and enables rapid responses when there are calls for proposals. Through dialogue common interests and contacts are recalled, and connections can be made across normal subject boundaries. These boundaries are differently located in each Member State, offering rich rewards for the intellectual traveller. Networks represent potential, whereas coalitions constitute a form of action. To the extent that Work Life 2000 has a programmatic agenda, it is assembling an international coalition of partner institutions with shared interests in at least discussing that agenda.

The way forward includes learning from differences. There is a rich resource of European experience, typically gained at the national level but with lessons applicable across the EU. The applications are not direct, but emerge in dialogue. In the Work Life 2000 workshops, the first day tends to be spent developing mutual acquaintance and trust, providing the basis for more detailed exploration and exchanges on the second and subsequent days. In general we find consistency in overall objectives between Europeans in a given profession, complicated by the variations in their political, economic and institutional circumstances. Simple generalisations break down, but within the context of emerging shared forms of life, understanding can develop. This is an important conclusion for the future of Europe, and it is borne out at each workshop. There is a European context which accommodates different and diverse views.

The New Organisation of Working Life Research

The organisation of academic research and governmental work is not immune to change, and new patterns and partnerships are emerging. Traditional research institutes have been maintained in most European Member States, with ongoing projects concerning working life. As the European economy harmonises and converges, this raises questions for the scientific research community.

- How can the results of research be disseminated to produce practical results?
- How can the agenda for working life research be attuned to the emerging European situation?
- How can the boundaries between the scientific disciplines be adjusted to meet changing needs?
- How can the different professionals concerned with working life accommodate to changing circumstances in the workplace and the European Union?

The Dublin Foundation and the Bilbao Agency provide potential bases for a European programme, and have been active partners in Work Life 2000.

The International Context

Within the European Union, no one Member State has the resources to sustain a large-scale research and development programme across the full range of subjects involved in working life. The first year of Work Life 2000 was also the first year of existence of the new UK Work Organisation Network, and each casts light on the other. We can learn from the differences as we look to the future.

Sweden has maintained a core research community concerned with working life, although the institutional structures have been reformed in recent years. The orientation of the research and dissemination has previously been Swedish and Nordic, but is now European, with the visible focus of the Swedish Presidency of the European Union in 2001. By contrast the United Kingdom has reduced both research and industrial relations structures over the past two decades. Uniquely in Europe, there is no national institute concerned with working life, although residual expertise from previous structures is still available. The recently elected government is committed to a central role in Europe although, like Sweden, they have declined to join the first stage of the single currency. Sweden comes to the new Millennium with a background of a strong role for the state, supported by a broad consensus that has lasted for most of the century. By contrast the United Kingdom is moving away from a fundamental reliance on market forces, as some of the social and economic consequences become more apparent.

These two Member States are typical of others in the European Union in needing a third way, based on interdependence rather than unaffordable independence. The Work Life 2000 series of workshops has laid a foundation for European Social Dialogue, bringing together experiences researchers, practitioners and social partners. Held largely, but not exclusively, in Brussels, it has begun to impact on the thinking of European Commission staff, who have found a valuable forum in which to test and develop ideas. Contributors from outside the European Union have noted the development of a constructive ferment of ideas within general shared values and objectives.

Success in the workplace, and in working life research, depends on blending theory and practice. Both Scandinavians and British have contributions to make: the work of the British Tavistock Institute was developed and extended in sustained national programmes in Scandinavia. This led to a wealth of reflection on workplace development. Interestingly, some of the models, based on networks and coalition approaches to organisational development, are applicable to the British situation, where new structures are required to fill a vacuum, drawing strength from the maturity of Swedish institutions.

Each European Union Member State has a slightly different story to tell, as we have learned in the early stages of Work Life 2000. The Irish have a similar history to the British in institutional terms, but with greater experience of deriving benefits from European membership. The Norwegians voted to stay outside the European Union, but are natural contributors to the Nordic research tradition, continuing to work with Sweden, Denmark and Finland. Finland are recent EU members, but with a real commitment to working life research, heightened by their awareness of the practical significance of demographic change. Denmark combines an approach to health service provision on the British model with a Scandinavian concern for working life. Germany has been proud of national programmes in the humanisation of work and technology, and moves from the achievement of reunification of Germany to an enthusiasm for harmonisation in more general terms. Austria are recent members, seeking to develop a distinctive role after their first experience of EU Presidency. The Netherlands in recent years have developed private sector solutions for many problems previously handled by the state, and tend to resist grand schemes. The Belgians, and to a lesser extent Luxembourg, have developed research and expertise in working life, learning from experience as multilingual states which host European institutions. France has had a central role for the state in the past, but is seeking to learn from the experience of partners in developing new institutions. Italy has retained many traditional approaches to work and institutions, but is aware of the emerging differences between Italian regions, and the lessons to be learned from European differences. Spain has benefited from European membership, and sees the coming years of expansion as critical. Portugal and Greece have been slow to be involved in working life research and debate, although the Greeks, in particular, locate the dialogue in a longer democratic tradition.

Readers of the first *Work Life 2000 Yearbook* who use the index of subjects will be able to trace the complexities of international debate and the interconnections of previously separate subjects.

The Organisation of Work Life 2000

The Swedish input to Work Life 2000 is worthy of particular mention. Working life has been a central concern to Swedish governments for decades, and the workshops are preparing the way for a European Conference, to reflect a European agenda. Most workshops have been led by Swedish researchers, who set the agendas and invited overseas participants, without seeking to dominate the proceedings. Keynote speakers came from all over the world, intrigued by the attractive agenda to which they were invited to contribute.

It was realised that the workshops had enormous potential in facilitating the European Social Dialogue. Accordingly popular summaries are produced following each workshop, as well as academic reports edited by the workshop leaders. The Yearbooks are intended to serve the growing community in dialogue, enabling participants in particular workshops to locate their discussions in a wider context. As rapporteur I seek to attend and document each of the workshops in the expanding series. The emerging agenda is indeed European, and the dialogues are conducted in European English, which uses English words but expresses concepts, such as social partnership, that have been alien to the British tradition.

After the success of the first year, the pace of workshops accelerates, organised in the same overall themes of:

1. Labour Market

2. Work Organisation

3. Work Environment

4. Small and Medium Sized Enterprises

5. Information Society

Reports and policy proposals will increase in frequency, together with the editing and publication of two further Yearbooks, culminating in the Work Life 2000 conference in Malmö in January 2001.

Indexes

Subject Index

Name Index

This CD-ROM contains the full text of
Work Life 2000 Yearbook 1 – 1999
Edited by Richard Ennals

The text is in a single file, called **WL2000_1.PDF**

If your system already has *Acrobat Reader* installed, you can usually see the text simply by clicking (or double-clicking, according to your system) on the file name.

Alternatively, open your *Acrobat Reader*, click on FILE, OPEN, then type the file location and address (such as D:\ WL2000_1) before pressing ENTER.

You can use *Acrobat Reader's* features to locate words or text (use TOOLS, FIND) or to print or copy parts of the text.

All text is © 1999 Swedish National Institute for Working Life.

If you do not have Acrobat Reader on your system, this CD-ROM includes all you need to install it on your IBM-compatible PC, running under Windows 95/98 or Windows 3.x

Two versions of *Acrobat Reader v.3* are included on this CD-ROM. They are located in the **Acrobat** directory. One is a 16-bit version for Windows 3.1, the other is a 32-bit version for Windows 95/98. The reader takes up about 5MB of disk space.

Windows 95/98:

Make sure all Windows applications are shut down. Use Windows Explorer to explore the sub-folder on the CD-ROM ..**Acrobat\32Bit** . Double-click on the file **Ar32e30** to begin installation. Follow the on-screen instructions.

Windows 3.1:

Make sure all Windows applications are shut down. Use File Manager to open the CD-ROM subdirectory ..**acrobat\16bit** . Double-click on the file **ar16e30.exe** to begin installation. Follow the on-screen instructions.